TRACE ELEMENTS IN PLANTS

TRACE ELEMENTS IN PLANTS

BY

WALTER STILES

M.A., Sc.D., F.L.S., F.R.S.

*Emeritus Professor of Botany in the
University of Birmingham*

CAMBRIDGE

AT THE UNIVERSITY PRESS

1961

CAMBRIDGE UNIVERSITY PRESS
Cambridge, New York, Melbourne, Madrid, Cape Town,
Singapore, São Paulo, Delhi, Mexico City

Cambridge University Press
The Edinburgh Building, Cambridge CB2 8RU, UK

Published in the United States of America by Cambridge University Press, New York

www.cambridge.org
Information on this title: www.cambridge.org/9781107698376

First edition 1946
Second edition 1951
Third edition 1961
First published 1961
First paperback edition 2013

A catalogue record for this publication is available from the British Library

ISBN 978-1-107-69837-6 Paperback

To

EDWARD JAMES SALISBURY

Director of the Royal Botanic Gardens, Kew
Chairman of the Mineral Deficiencies Conference of the
Agricultural Research Council

IN FRIENDSHIP

CONTENTS

viii CONTENTS

PREFACE TO THE THIRD EDITION

In the preparation of the new edition of this book considerable alterations have been made to the original text. In the first place, as the title indicates, the subject-matter has been limited to a discussion of trace elements in plants, the effects of trace-element deficiency or excess in animals being considered only where such effects arise directly from shortage or excess of these elements in the plants eaten by grazing animals. To have included a full discussion of trace elements in animals would have meant unduly increasing the size of this book; moreover the subject would be dealt with much more satisfactorily by a zoologist than a botanist.

A very great deal of fresh information about trace elements in plants has accumulated during the last sixteen years, to deal adequately with which has involved many additions to, and much rearrangement of, the text. In this new edition some consideration has been given to the results of excess absorption as well as to those of deficiency; the consequences of excess of a trace element may be as serious as those of its shortage. Some reference has also been made to soil conditions as they relate to availability of the trace elements.

The increase in size necessitated by additions to our knowledge has been lessened to some extent by the deletion of matter relating to trace elements in animals and by the removal of the lists of species for which various trace elements have been proved essential. Since there is now every reason to believe that the five universally recognized trace elements are essential for all higher plants, there is little point in listing those, now very numerous, in which essentiality of one or other of the trace elements has actually been demonstrated.

It is a pleasure to me to acknowledge with gratitude the very great help I have received from Dr E. J. Hewitt in providing a number of excellent photographs obtained at Long Ashton

Research Station where the investigation of trace element problems has been in progress for many years. It is to be noted that the copyright of these photographs (Plates 3, 4, 10–14 inclusive, 16 A and B) remains the property of Long Ashton Research Station. My thanks are also due to Messrs M. Greenwood and R. J, Hayfron and the Oxford University Press for permission to use the photograph showing zinc deficiency in cacao (Plate 5 B). W.S.

READING
26 February 1960

PREFACE TO THE FIRST EDITION

T RAC E elements, micro-nutrients and minor elements are terms applied to a number of chemical elements which are essential for the lives of plants and animals, but which are required in extremely small quantity. The development of our knowledge of the part played by the trace elements in the life of plants and animals is very recent, nearly all our present information on the subject having been acquired during little more than the last twenty years. Nevertheless, in the course of that time a great many observations have been made in the field, and much experimental work carried out in both field and laboratory, while many hundreds of publications, some of slight value, others of considerable scientific and economic importance, dealing with trace elements in living plants and animals have appeared in the scientific press. The time thus seems ripe for the presentation of a digest of this information, so that the salient facts may be available in a convenient form, and the present position of our knowledge of trace elements in living organisms made plain. It is with this intention that the present book has been written.

The role of the trace elements in organisms is, in the first place, a matter for the plant or animal physiologist, but as deficiency or excess in the supply of the various trace elements may lead to a diseased condition of the plant or animal with serious economic consequences, the trace elements are also of interest to the plant pathologist, the veterinary surgeon, the agriculturist and horticulturist. Indeed, as regards plants at any rate, much more is definitely known of the plant pathological aspect of the trace elements than of their physiological functions.

I would take this opportunity of expressing my thanks to Mr W. Morley Davis of Harper Adams Agricultural College, who first introduced to me the effects of trace-element deficiencies in the field, and who has generously and constantly put at my disposal both his knowledge of the pathological effects of trace-element deficiency, and the pathological material itself. I also have pleasure in thanking Dr C. S. Piper of the Waite Agricultural Research Institute, University of Adelaide, for permission to use the photographs reproduced in Plates 2 and 9, and my colleagues Dr K. W. Dent and Mr A. E. Roberts for the preparation of the rest of the photographs in the book.

WALTER STILES

BIRMINGHAM
28 October 1944

ON the occasion of the reprinting of this book I have taken the opportunity to make a few minor corrections and additions to the text and to add a photograph showing the effect of molybdenum deficiency in cauliflower. For permission to use this photograph my thanks are due to Messrs E. J. Hewitt and E. W. Jones of the Long Ashton Research Station and to the Editors of the *Journal of Pomology and Horticultural Science*.

W.S.

BIRMINGHAM
28 July 1948

PLATES

LIST OF ABBREVIATIONS

Throughout the book the following contractions have been used:

ml (millilitre)   = one-thousandth of a litre
μg (microgram)   = one-millionth of a gram
Å (Ångström unit) = one ten-millionth of a millimetre
p.p.m.   = parts per million

CHAPTER I

HISTORICAL INTRODUCTION

In the year 1699 Woodward published the results of experiments in which cuttings of plants were grown not in soil but in rain water. More than a century and a half later Julius Sachs (1860) and W. Knop (1860) independently developed this method of water culture by growing plants of several different species in a dilute aqueous solution of various salts. In this way the materials available for absorption by the roots of the growing plants were controlled, and Sachs and Knop concluded from their experiments that so long as the culture solution contained salts involving the elements nitrogen, sulphur, phosphorus, potassium, calcium, magnesium and iron, a perfectly healthy plant would result. The elements occur in soil as constituents of compounds present in, or derived from, the minerals of the underlying rock, and are therefore generally known as mineral nutrients. Experiments of this kind have been repeatedly carried out by subsequent investigators and many formulae for water-culture solutions have been used and recommended. The water-culture method has also been extensively used for research in plant nutrition. Some of the best known and most widely used water-culture solutions are those of Sachs and Knop themselves and the later ones of Pfeffer and Von der Crone. Plants of a great number of species have been successfully grown in water culture, and the method has been advocated, perhaps too optimistically (cf. Hoagland and Arnon, 1938), as a means of cultivating certain crop plants on a commercial or semi-commercial scale. The compositions of some of the best known and most widely used water cultures are given in Tables 1–3.

TABLE 1. *Sachs's nutrient solution*

Potassium nitrate	1·0 g
Sodium chloride	0·5 g
Calcium sulphate	0·5 g
Magnesium sulphate	0·5 g
Calcium phosphate	0·5 g
Water	1 l.

A few drops of a solution of ferric chloride or ferrous sulphate.

TABLE 2. *Pfeffer's stronger nutrient solution*

Calcium nitrate	4·0 g
Potassium nitrate	1·0 g
Magnesium sulphate	1·0 g
Potassium dihydrogen phosphate	1·0 g
Potassium chloride	0·5 g
Water	3 l.
Ferric chloride medicinal solution	3 to 6 drops

Pfeffer's weaker solution is similar, except that the same quantity of these salts is dissolved in 7 l. of water.

TABLE 3. *Von der Crone's nutrient solution*

Potassium nitrate	1·0 g
Calcium sulphate	0·5 g
Magnesium sulphate	0·5 g
Calcium phosphate	0·25 g
Ferrous phosphate	0·25 g
Water	1 l.

It will be noticed that, in addition to the elements given above, Pfeffer's solution also contains chlorine, while Sachs's solution contains not only chlorine, but sodium. While the general opinion of plant physiologists for many years was that most plants required neither of these elements, it was rather vaguely held that some plants required sodium or chlorine in addition to those regarded as always necessary. Thus Pfeffer (1900) remarked that probably no plant had been grown in complete absence of chlorine, and he considered it uncertain whether a minimum amount was essential or not. With regard to sodium, not only did Osterhout (1912) consider this element to be essential for marine algae, but from experiments on the growth in length of the roots of seedlings of wheat and other plants (1908) he decided that sodium had some definite function in the growth of these plants. Several workers have recorded increased growth of plants as a result of the addition of sodium to the medium.

So opinion on the mineral nutrition of plants remained during the latter half of the nineteenth century. It was well established that many elements, apart from those regarded as necessary, were present in plants, but their presence was regarded as incidental, due to their presence in the soil in which the plants grew. But in 1897 G. Bertrand claimed that manganese was constantly associated in the plant with an oxidizing enzyme, laccase, and he came to regard manganese as an essential constituent of the oxidase system. In 1905 he claimed that an insufficient supply of manganese to plants brought about diminution or cessation of growth, and that manganese was to be regarded as an essential element. Much subsequent experimental work and many field observations have gone to confirm Bertrand's view.

In 1914, by the use of water-culture experiments with carefully prepared culture vessels and nutrient salts, Mazé showed that not only manganese but zinc also was necessary for the growth of maize. As a result of later work on similar lines he claimed that aluminium, boron, chlorine and silicon were all essential in small amounts for the healthy growth of plants. Of these various elements, boron had already been found by Agulhon (1910) to induce an increased production of dry matter in wheat, oats and radish grown in sand cultures and in maize, colza and turnip grown in field-plot experiments. It is, however, scarcely correct, as has sometimes been done, to quote Agulhon as the first worker to point out the *essential* nature of boron for higher plants. As Dennis and O'Brien (1937) have pointed out, Agulhon provided no proof that normal growth was not possible without boron, and the credit for first calling attention to the necessity of this element for any plant is due to Mazé. The importance of boron as an essential plant nutrient was brought into prominence in 1923 by Miss Warington, who showed by means of water cultures that while a concentration of boric acid as low as 1 in $12 \cdot 5 \times 10^6$ was sufficient to allow the normal growth of the broad bean (*Vicia faba*), in complete absence of boron death of the plant supervened after the development of quite characteristic pathological symptoms. A few years later Sommer and Lipman (1926) added cotton, castor oil, buckwheat, flax, mustard and barley to the list of plants for the growth of which a supply of boron is essential. Since then the list of such species has been very greatly extended.

Confirmation of Mazé's finding that zinc is necessary for the healthy development of maize was provided by the observations of Barnette and Warner (1935), Mowry and Camp (1934) and others, who showed that the curious chlorotic condition of this plant known as 'white-bud' could be cured by application of zinc sulphate. Sommer and Lipman (1926) also found zinc essential for the growth of sunflower and barley, and later Sommer (1928) reported that buckwheat and beans could only undergo normal development in presence of zinc. In beans, for example, abscission of leaves and flower buds occurred in cultures deprived of zinc, whereas in controls adequately supplied with this element flowering took place and seeds were produced. A number of trees have been shown to require a supply of this element.

As regards the remaining elements that Mazé concluded were necessary for the growth of maize, reference has already been made to the doubt which has existed for many years about the necessity of chlorine for plant development. As long ago as 1862, Nobbe and Siegert concluded that chlorine was essential for the normal development of buckwheat, but many subsequent experiments carried out to check this conclusion yielded conflicting results. But in 1938 the much-debated question of the effect of chlorine on the growth of buckwheat was examined by Lipman, and it was found that plants grown with 5 p.p.m. of chlorine added as potassium chloride produced markedly more dry matter and seeds than plants grown without added chlorine. Moreover, the seeds from plants grown with added chlorine gave a higher percentage of germination than the seeds from plants without added chlorine, and of the seedlings produced from the latter little more than half developed to any extent. A second series of cultures was made from the seed obtained in the first series, the seedlings from seeds obtained from the cultures with added chlorine being again supplied with this element, and those from seeds grown in cultures without added chlorine being again deprived of it. The superiority of the plants supplied with chlorine was emphasized, and Lipman concluded that if chlorine was not absolutely essential to buckwheat it certainly greatly influenced its growth and seed production. A like conclusion was drawn from water-culture experiments with peas.

As regards aluminium, Stoklasa in 1922 grew a number of hydrophytes in water cultures and silica gel cultures with and without aluminium, and found growth was considerably improved in presence of aluminium. Indeed, silica gel cultures of *Glyceria aquatica* without aluminium died in 22 days and aluminium-free water cultures of *Juncus effusus* died in from 56 to 69 days. Addition of aluminium also improved the growth of wheat, barley and oats, but no beneficial effect of this element was observed with plants of a number of other species, including

Polygonatum officinale and *Iris bohemica.* More critical water-culture experiments to test the essentiality of aluminium were carried out by Sommer (1926) with peas and millet. Specially purified nutrient salts were used in making up the culture solutions, and in those containing aluminium this element was present to the extent of 5·5 or 1 p.p.m. of the solution. The presence of 1 p.p.m. brought about a slight increase in the dry weight of peas, but with millet the presence of aluminium brought about a definite increase in dry weight and a very pronounced increase in the weight of seeds produced, the respective weight of seeds produced with and without aluminium being 4·98 and 0·23 g.

In 1938 Lipman obtained similar results with sunflowers and maize, the effect being particularly noticeable with the latter, where the addition of aluminium sulphate to the extent of 1 p.p.m. of aluminium (renewed from time to time) resulted in an increase in dry weight of the vegetative parts of about 20 per cent and an increase in the dry weight of the ears of about 155 per cent.

Sommer and Lipman also investigated the essentiality of silicon. Sommer in 1926 found that the presence of silicon, added as silica gel, increased the dry weight of rice plants from 4·4 to 7 g, and she concluded that this increase was big enough to indicate that silicon is essential for the growth of rice. Marked increases in seed production as a result of the presence of silicon in the culture solution were observed with millet. Lipman in 1938 found that sunflowers and barley grown in water culture produced more dry matter in a given time with colloidal silica added to the solution than without such addition. With barley the addition of silica also resulted in a greater yield of grain.

Since Lipman could not be sure that in his control cultures without added silicon, aluminium or chlorine respectively, there was not a small quantity of the particular element present, he did not go further than emphasize the importance of silicon, aluminium and chlorine for the plants used and probably for higher plants in general. However, he regarded the probability of the indispensability of these elements as very great indeed. As regards chlorine at any rate more recent work has lent support to this conclusion, and has indicated that chlorine may be a more generally essential element for higher plants than was previously thought.

Convincing evidence that copper was essential for the normal growth of a number of plants was produced in 1931 by Sommer and by Lipman and Mackinney. Sommer found that the addition of small quantities of copper induced a considerable increase in growth of sunflowers, flax and tomatoes, while in the same year Lipman and Mackinney found that flax and barley grown in water culture failed to produce seed in complete absence of copper. Evidence was also presented by Anderssen (1932) in South Africa and by Oserkowsky and Thomas (1933) in America indicating that chlorosis and dying back of the branches of various fruit trees result from a deficiency of copper. Further, a pathological condition of cereals and some other plants known as 'reclamation disease', occurring in plants growing on reclaimed heath land in Holland and elsewhere, was attributed by Sjollema in 1933 to copper deficiency. In cereals exhibiting this condition the tips of the leaves turn yellow or white while seed fails to form. Plants of so many species have now been shown to require a small supply of copper that it is reasonable to suppose that this element, like manganese, zinc and boron, is generally necessary for the growth of higher plants.

In 1939 Arnon and Stout obtained evidence that molybdenum might be an essential element for higher plants. They found that tomato plants exhibited pathological symptoms when grown in a culture solution containing all the ordinary nutrient elements along with manganese, boron, zinc and copper but which was freed from all trace of molybdenum. The lower leaves of the plants first developed a mottling; then dying of the marginal cells followed while the flowers fell without setting fruit. The condition was prevented by the addition of one part of molybdenum as molybdic acid in 100 000 000 to the culture solution, while the pathological symptoms were removed in the molybdenum-deficient plants by spraying the leaves with a very dilute solution of molybdic acid. Subsequently, plants of a number of other species have been shown to require a supply of molybdenum, and although the possibility of exceptions is not to be ruled out, there is every reason to regard molybdenum as an essential element for the growth of many higher plants. Molybdenum was reported to be necessary for the optimum growth of peas by Bobko and Savvina in 1940.

In 1941 Steinberg concluded that not only molybdenum, but also gallium is an essential element for the growth of *Lemna*. Since gallium has properties resembling those of aluminium and may occur as an impurity in preparations of aluminium salts, Steinberg suggested that the effect on plants attributed to aluminium might in fact be due to gallium present as an impurity in the aluminium compounds used in the preparation of the nutrient medium.

So far only higher plants have been considered, but it is quite clear that the same trace elements may be equally essential for the growth of lower plants, at any rate of fungi. Indeed, the necessity of an element for the growth of a fungus has in more than one instance been demonstrated before its essentiality for a higher plant has been claimed.

The favourite species among the fungi for studies on mineral nutrition has been *Aspergillus niger*. In addition to the generally acknowledged mineral nutritive elements, with the exception of calcium,* but definitely including iron,† it has been claimed that this fungus definitely requires zinc, manganese, copper, molybdenum, gallium and vanadium. Indeed, as long ago as 1869, Raulin showed that for *A. niger* zinc is necessary, although the significance of this finding was rather obscured subsequently by the attempt to explain the increased growth resulting from the presence of a small quantity of zinc as a stimulation effect on normal development due to the action of a poison in low concentration. While the possibility of such a stimulation effect in general is not to be ruled out without proof, such proof is forthcoming when it has been shown that with rigid exclusion of any

---

* There exists considerable doubt about the necessity or otherwise of calcium for the growth of fungi. The general opinion at present appears to be that this element is necessary for the growth of some species but not of others (cf. Steinberg, 1948). Much experimental work is necessary to place our knowledge of this question on a reliable basis. According to Davis, Marloth and Bishop (1928) calcium is necessary for the development of *A. niger*. Mann (1932), on the other hand, came to the opposite conclusion.

† Iron is placed in the category of micro-nutrients by some workers on fungus nutrition, e.g. Steinberg (1939) and Foster (1939).

substance normal development is prevented. For zinc this proof was definitely provided by Steinberg (1919), who, by means of a special technique, succeeded in removing all but the minutest traces of zinc from the culture medium, and then showed that the cultures of *A. niger* provided with zinc produced more than 2000 times as much dry matter as the controls deprived of all but the last traces of that element. Steinberg's result was later confirmed by Bortels (1927), Roberg (1928, 1931) and Gollmick (1936).

The need for manganese for the normal development of *A. niger* was claimed by Bertrand and Javillier (1911a), and this was confirmed by Steinberg (1935a). That copper is essential for the growth of this fungus was shown by Bortels (1927) and by Wolff and Emmerie (1930). The rigid proof of the need of molybdenum for the growth of *A. niger* was given by Steinberg in 1937, while in the following year the same worker added gallium to the list of elements essential for this fungus. D. Bertrand (1941) considered vanadium played an essential part in the growth of *A. niger*. This conclusion was based on the finding that addition of a small quantity of vanadium to the culture medium resulted in an increase in growth of 20·9 per cent over that of the control grown on the same medium without added vanadium. Analysis of the controls showed that they contained a certain amount of vanadium in spite of the freedom of the culture medium from this element and it would seem possible that this vanadium had been obtained from the culture vessels.

While the mineral nutrition of no other fungus has received so much attention as that of *A. niger*, sufficient information has now accumulated to justify the conclusion that for the fungi in general a supply of trace elements is necessary. Thus McHargue and Calfee (1931) concluded that zinc, manganese and copper are necessary for the growth of *A. flavus* and *Rhizopus nigricans*, and the same three elements were considered necessary for *Ceratostomella ulmi* by Ledeboer (1934), for *Trichophyton interdigitale* by Mosher, Saunders, Kingery and Williams (1936), and for *Phymatotrichum omnivorum* by Rogers (1938). Foster and Waksman (1939) reported that *Rhizopus nigricans* failed to produce zygospores in absence of zinc, while the favourable effect on development produced by zinc and copper on a number

HISTORICAL INTRODUCTION 9

of fungi belonging to different families was reported by Metz (1930) and for a number of species of *Aspergillus* by Roberg (1928, 1931). In 1933 Lockwood had reported that the growth of *Penicillium javanicum* was increased by additions to the medium of molybdenum as well as columbium and tungsten. Thus, as far as our information goes at present, both higher plants and fungi, or some of them, require a supply of manganese, zinc, copper and molybdenum. Many higher plants have been shown to need boron, but little attention has been given to this element by workers on fungus nutrition. In 1928 Davis, Marloth and Bishop reported that the yield of a species of *Dothiorella* grown on an artificial medium was reduced to half by the removal of traces of boron from the nutrient salts used but Blank (1941) could find no indication that boron was essential for the growth of the cotton root rot fungus *Phymatotrichum omnivorum*.

Not a great deal of information is available about the trace-element requirements of plants other than angiosperms and fungi. Among conifers Kessell and Stoate (1938) found zinc essential for *Pinus radiata* and among the Pteridophyta Hunter (1953) recorded the development of chlorosis in bracken (*Pteridium aquilinum*) as a result of manganese deficiency. As regards algae manganese has been shown to be essential for the unicellular green alga *Chlorella* by Hopkins (1930a, b, 1934) and for the diatom *Ditylum brightwelli* by Harvey (1939). Roberg (1932) reported increased growth of two unicellular green algae *Coccomyxa simplex* and *Chlorella vulgaris* as a result of small additions of salts or iron, zinc and copper to the normal nutrient solution. From his work, however, it was not made clear that zinc and copper were actually essential for the growth of these plants, and Roberg himself said that these elements were to be considered as acting as stimulants. However, it seems clear that the same trace elements normally required by higher plants are also essential for algae. Thus the need for copper and molybdenum by *Chlorella pyrenoidosa* was established by Walker (1953) who also (1954) confirmed its need of manganese and zinc, while as early as 1935 Geigel had reported that the presence of boron in a concentration of 10 p.p.m. brought about a marked stimulation of the growth of *Chlorella*.

As regards the essentiality of molybdenum Arnon, Ichioka, Wessel, Fujiwara and Wolley (1955) found that when *Scenedesmus obliquus* was grown in a nutrient medium in which the nitrogen was provided exclusively as nitrate a small amount of molybdenum was essential for growth and cell division. Other observed results of molybdenum deficiency were a marked reduction in chlorophyll content, an accumulation of starch in the cells and, indeed, a general appearance similar to that resulting from nitrogen starvation. It was, however, shown by Ichioka and Arnon (1955) that the necessity for molybdenum was removed if nitrogen was supplied as either ammonium carbonate or urea.

Arnon and Wessel (1953) concluded that vanadium was an essential element for the growth of *Scenedesmus obliquus* as this alga showed a marked increase in growth as a result of adding vanadium in a concentration of 0·01 p.p.m. to the culture solution.

Silicon was shown by Lewin (1954, 1955) to be an essential constituent of the diatom *Navicula pelliculosa* and it may reasonably be concluded that this is so for the whole group.

Some interesting information has been obtained regarding trace-element requirements of the blue-green algae. Bortels, who in 1930 had shown that molybdenum was necessary for the fixation of nitrogen by the nitrogen-fixing bacterium *Azotobacter chroococcum*, showed later (1940) that molybdenum played a similar part in various nitrogen-fixing blue-green algae. However, in these as well as in *Azotobacter*, he found that the molybdenum could be partially replaced by vanadium.

Although cobalt is known to be an essential element in the nutrition of animals, no evidence has so far been obtained to suggest that it is necessary for the growth of higher plants. Holm-Hansen, Gerloff and Skoog (1954) have, however, claimed that cobalt is essential for four species of blue-green algae they examined. These were two nitrogen-fixing species, *Nostoc muscorum*, and *Calothrix parietina*, and two non-nitrogen-fixing species, *Coccochloris peniocystis* and *Diplocystis aeruginosa*. Deficiency of cobalt was found to result in reduced growth and the development of a chlorotic condition, symptoms which disappeared on addition of cobalt. Maximum rate of growth of the algae was obtained with cobalt in the very low concentration of

0·40 μg per litre or 1 in 2500 million, while a definitely beneficial effect was noted with a concentration 1/200th of this.

The observations of Kratz and Myers (1955) indicated that sodium also might be essential for the growth of blue-green algae. With three species, *Nostoc muscorum*, *Anabaena variabilis* and *Anacystis nidulens*, they found that both sodium and potassium were necessary if maximum growth of these plants was to be maintained. Deficiency of sodium resulted in a lessening of the rate of growth below the maximum which was re-established on supplying sodium.

From what has so far been discovered regarding the essentiality of trace elements, two questions arise which only further research can answer. The first is how far the necessity for these elements is general throughout the plant kingdom, and the second is whether we now know the complete list of these elements. As regards the first question, it would seem probable rather than merely possible that when plants differing as widely, both taxonomically and anatomically, as well as physiologically, as angiosperms and fungi both exhibit these requirements, we are dealing with something very fundamental in plant nutrition, and we are justified in concluding that the best established of the trace elements, manganese, boron, zinc, copper and molybdenum, are likely to be found essential for the nutrition of plants in general. In regard to the second question, Steinberg (1938c) has contributed an interesting discussion on the relations between essentiality of elements and their atomic structure, and he draws the conclusion from such considerations that scandium may be an essential element for plant nutrition. In this connexion some experiments carried out by Arnon are of interest. In 1937 this worker described the results of water-culture experiments in which the growth of barley plants was improved by the addition of small quantities of molybdenum, chromium and nickel. In the following year further experiments were described in which plants of asparagus and lettuce were grown in four different culture solutions. The first of these contained the ordinary nutrient elements. The second contained these together with the four well-established micro-nutrients, manganese, boron, zinc and copper (designated A 4) and some chlorine. The third contained all the elements present in the second solution together with another seven (designated B 7); these were molybdenum, titanium, vanadium, chromium, tungsten, cobalt and nickel. The fourth solution contained all the elements present in the third solution together with thirteen other elements, namely, aluminium, arsenic, cadmium, strontium, mercury, lead, lithium, rubidium, bromine, iodine, fluorine, selenium and beryllium. This group was designated C 13. The fourth solution also contained sodium. All the elements of the A 4, B 7 and C 13 groups were present in very low concentration, that is, as traces. The four solutions, which may be denoted by I, II, III and IV, thus contained respectively the following mineral elements:

    I: N, S, P, K, Ca, Mg, Fe.
   II: elements of I + B, Zn, Mn, Cu, + Cl.
  III: elements of II + Mo, Ti, V, Cr, W, Co, Ni.
   IV: elements of III + Al, As, Cd, Sr, Hg, Pb, Li, Rb, Br, I, F, Se, Be,
       + Na.

The fresh weights in grams of the plants grown in these different solutions are shown in Table 4. Thus the great effect of the four well-established micro-nutrients is well demonstrated, but the seven additional elements of solution III produced a further increase in growth of asparagus, while for lettuce their effect was most striking, the yield being increased about ten times as a result of their addition. The subsequent demonstration by Arnon and Stout that molybdenum is necessary for the tomato has been mentioned earlier, and this leaves in doubt whether the effectiveness of the seven elements of solution III not included in solution II is due solely to the molybdenum or whether some or all of the others are also in part responsible for this. That no further increase in yield results from the addition of the extra thirteen elements of solution IV at first sight suggests that none of these is essential for the growth of asparagus and lettuce, but it may be that one or more of these is actually necessary but exists in sufficient quantity as impurity of one of the salts used in making up solution III.

Using Arnon's technique, Twyman (1943) obtained a similar result with oats. Four weeks after germination of the grains the average dry weights of plants supplied with the A 4, B 7 and A 4 + B 7 groups of trace elements were respectively 0·101, 0·130 and 0·238 g. This again showed the necessity of one or more of the elements of Arnon's group B 7. With the establishment of molybdenum as in all probability a generally essential element in plant nutrition, in later work with groups of trace elements molybdenum was transferred to the A group, which thus became A 5 while the B group became B 6.

TABLE 4. *Growth of plants of asparagus and lettuce in four different culture solutions.* (Data from Arnon)

	Fresh weight (g)			
Culture solution	Asparagus		Lettuce	
	Shoots	Roots	Shoots	Roots
I	16·2	12·8	71·4	14·5
II	88·2	38·2	105·7	22·0
III	118·1	81·3	1068·3	188·6
IV	121·7	74·5	984·4	196·2

The question arises why for so many years the necessity for the trace elements in plant nutrition was not recognized. The answer given to this question by Mazé was no doubt the correct one. Mazé considered that the importance of the micro-nutrients had been overlooked because in water-culture experiments a sufficient quantity of these was introduced into the cultures from (1) the seeds used, (2) impurities in the salts used in preparing the culture solutions, and (3) solution from the vessels containing the culture solutions. Indeed, knowledge of the existence of the various trace elements has only been obtained through the purification of the water and nutrient salts used, and the choice of suitable culture vessels.

The securing of an adequate degree of purity of the materials used is of the utmost importance in experimental work designed to examine the indispensability or otherwise of particular substances. The methods that have been developed to this end, along with other experimental methods of value in work on micro-nutrients, form the subject of the next chapter.

Although addition of a compound of a particular element to the nutrient medium in which plants are growing may bring about increase in rate of growth of the plants, it does not follow that the element is essential for the growth of plants. Indeed, increases in growth rate as a result of the addition of compounds of a number of different elements have been recorded from time to time.

Among observations of this kind particular mention may be made of those of R. S. Young (1935), who examined the effect on the growth of timothy (*Phleum pratense*) of thirty-five of the rarer elements when added in five different concentrations (2000, 500, 100, 10 and 0·1 p.p.m.) to a sandy loam. Beneficial effects were observed with molybdenum, supplied in 2000 p.p.m., and with antimony, barium, bismuth, bromine, cerium, manganese, strontium, tungsten, uranium and yttrium, supplied at the rate of 500 p.p.m. Aluminium, cadmium, copper, fluorine, lanthanum, lead, mercury, tin and zinc gave an increased growth when supplied at the rate of 100 p.p.m., while with 10 p.p.m. arsenic, beryllium, chromium, iodine, lithium, selenium, thorium, titanium, vanadium and zirconium were beneficial. At a concentration of 0·1 p.p.m., boron, nickel and thallium brought about an increase in growth. Of the thirty-five elements, the effects of which were tested, only cobalt appeared to be slightly toxic at this lowest concentration employed, while silver at this concentration appeared to have no effect. At higher concentrations than those stated for the respective elements the action of these was depressing on growth.

Experiments with cultures of two green algae, species of *Chlorella* and *Crucigina* respectively, led Young to conclude that on the whole any element will stimulate the growth of algae at a definite concentration which depends on the element and the species.

We certainly cannot conclude that an element which stimulates growth is necessarily essential for growth, although if increased growth is observed to result from the presence of a particular element there is always the possibility that that element may be an essential one. Whether it is so or not can only be proved by growing the plant in carefully controlled cultures, in which every care is taken rigidly to exclude from the culture medium the element under examination.

The necessity of a supply of manganese, zinc, boron and copper has now been demonstrated for so many different species that in absence of any evidence to the contrary it may with reason be concluded that these elements are all essential for the growth of vascular plants and algae. The same would appear to be true of fungi with the qualification that the status of boron in this respect is not so clear. As regards molybdenum there is every reason to believe that this is essential for plant growth when the source of nitrogen is nitrate or elemental nitrogen and it would seem likely that for many plants it is generally essential whatever the nitrogen source. Apart from these five well-established trace elements there is now evidence for the essentiality of chlorine, sodium, silicon, aluminium and gallium for certain species. Whether any of these are actually necessary for all plants only future work can decide.

In addition to the elements mentioned above claims have been made that many others, in appropriate concentration, 'stimulate' plant growth, and it is possible that for many elements there is a range of concentrations over which that element induces an increased rate of growth while in higher concentrations the effect is a retardation of growth, which is severer with increasing concentration. Such a reaction to the presence of the element does not prove that the element is essential for growth. Essentiality is proved only when, with rigid exclusion of the element from the nutrient medium, growth is either completely inhibited or reduced so drastically that what does occur can be attributed to traces of the element which the most careful control has failed to exclude.

CHAPTER II

METHODS OF INVESTIGATING MICRO-NUTRIENT PROBLEMS

1. The Purification of Materials used in Culture Experiments

The existence of the micro-nutrients raises a number of problems which can only be solved after the development of methods designed specially to deal with them. In the first place the question inevitably arises as to how we can be certain that any particular element is really essential or not. This is obviously a problem of mineral nutrition requiring immediate solution; it is clearly an essential preliminary for all investigations on the micro-nutrients that we should know what these are.

It has already been pointed out that the essentiality of the micro-nutrients for plant development was overlooked for half a century simply because an adequate supply of them was introduced into the culture (1) from the seed from which the plant developed, (2) from impurities present in the water and salts used in preparing the culture solutions, and (3) by solution from the vessels used to hold the culture solutions. Hence the problem of determining the micro-nutrients clearly resolves itself into devising means for preventing the introduction of micro-nutrients from these three sources. If this can be done the effect of the absence or presence of any particular element in the culture medium can then be determined for a wide range of plant species.

The prevention of the introduction of any particular elements from the seed used for the cultures is perhaps scarcely possible. But when the seed is small the amount of any micro-nutrient present in it is likely to be negligible, and when the seed is large the amount introduced can often be considerably reduced by removal of the cotyledons from the seedling as soon as the latter is established. Reference has already been made to Lipman's method of meeting the difficulty in his experiments with buck-wheat, in which seed was used from plants grown in culture solutions devoid of chlorine. It appears clear that by this procedure seed was obtained containing a negligibly small quantity of this element.

The problem of obtaining water and salts free from the trace elements for work on the nutrition of fungi has been dealt with by a number of workers, notably Steinberg (1919, 1935 b). It may be stated at once that pure salts sold as analytical reagents may contain a sufficiency of trace elements present as impurities to allow the growth of plants, nor does recrystallization necessarily afford an adequate means for the removal of these contaminants.

Steinberg's original procedure for obtaining a nutrient solution free from so-called heavy metal contaminants consisted in heating the complete nutrient solution with pure calcium carbonate under pressure. The nutrient solution used was one due to Pfeffer and was made up as shown in Table 5.

TABLE 5. *Pfeffer's nutrient solution*

Sucrose	50 g
Ammonium nitrate	10 g
Potassium dihydrogen phosphate	5 g
Magnesium sulphate	2·5 g
Ferrous sulphate	Trace

Since Steinberg's treatment of the solution leads to the removal of iron it is obvious that the trace of ferrous sulphate may be omitted.

To a litre of this solution 15 g of pure calcium carbonate were added and the mixture heated in an autoclave for 20 min under a pressure so as to give a temperature of 120·5 °C. The mixture was then allowed to stand overnight and the clear solution then decanted from the sediment. Subsequently Steinberg recommended filtering the solution from the sediment immediately after autoclaving.

The principle involved in this method of purification consists in increasing the alkalinity of the solution so that the traces of heavy metals are precipitated as carbonates, hydroxides and phosphates which are adsorbed on calcium salts precipitated in some bulk. In this way traces of iron, manganese, zinc and copper are removed from the solution.

Various modifications of this procedure have been proposed. Steinberg himself pointed out that the substitution of basic magnesium carbonate for calcium carbonate has certain advantages where work on fungi is concerned. This substitution avoids the introduction into the solution of calcium, an element which, as we have already seen, is of doubtful significance in fungal nutrition. Indeed, it appears to effect a pretty complete removal of any calcium present as impurity in the nutrient solution. Also the use of an autoclave is unnecessary, heating for 20 min at 100 °C being sufficient to precipitate the heavy metals. Care has, however, to be taken to avoid excess of the basic magnesium carbonate, as otherwise more or less complete removal of phosphate may result.

Bortels (1927) similarly purified the nutrient solution by precipitating the traces of the heavy metal contaminants with a small quantity of ammonium sulphide and adsorbing the precipitate on charcoal. Actually the ammonium sulphide appears to be unnecessary, according to Roberg (1928), who also purified the charcoal from mineral ash constituents by a preliminary treatment with acid. However, Steinberg pointed out that removal of the ash constituents appears to reduce the adsorbing power of charcoal, and he concluded that the use of charcoal is only advisable when for some reason it is essential to avoid the use of an alkaline earth compound.

In 1927 Hopkins and Wann made use of the adsorptive property of calcium phosphate for the removal of iron from culture solutions for the green alga *Chlorella* and later, for work with green algae and *Lemna*, Hopkins (1934) again employed calcium phosphate as an adsorbent for removal of traces of manganese from the nutrient solution. Sakamura, who had previously (1933, 1934) used charcoal for removal of traces of heavy metals from the nutrient medium of *Aspergillus* spp., concluded (1936), by polarographic examination* of nutrient solutions after treatment with charcoal and calcium phosphate respectively, that the latter effected a much more complete removal of the heavy metal contaminants, a conclusion which was confirmed by growth experiments. His procedure was as follows. Calcium phosphate was first purified by washing with water distilled in a glass still, a suspension of 50 g calcium phosphate in a litre of distilled water being shaken for 5 h, during which time the water was changed four times. The calcium phosphate was then filtered off

* See this chapter, p. 23.

with the use of ash-free filter paper and dried. Calcium phosphate so purified was then added to the culture solution so that it was contained in the latter to the extent of 0·5 per cent and the whole was brought to a pH of 5·5 by means of sodium hydroxide. The solution was then shaken for 2 h and twice filtered. The final filtrate comprised the working culture solution.

The preparation of culture solutions free from trace elements for work with higher plants presents a somewhat different problem from the preparation of nutrient solutions for the growth of fungi. It will be appreciated that it is impracticable to purify the complete culture solution for higher plants owing to the large quantities of solution required, and it is thus necessary to remove the trace elements from the water and nutrient salts separately. The means by which this may be done have been described in detail by Stout and Arnon (1939). As regards the water used, it is necessary to avoid the use of distillation apparatus made of or containing copper, silver, tin or other metal. Although the actual content of contaminants in distilled water from a metal still may be very small, yet the quantity of water used, particularly in the culture of higher plants, is very considerable, so that the absolute amount of contaminant presented to the plant may be far from negligible. Stout and Arnon found that ordinary distilled water contained from 0·1 to 0·01 p.p.m. of metal contaminants. They recommended, therefore, that water should be redistilled, using a trap and condenser of pyrex glass and distilling at a rate slow enough to give a cool distillate. Water is obtained in this way free from metal impurities. It should be noted that the use of Jena glass is to be avoided, since this contains zinc which may appear in the distilled water. For some work on trace elements it has been found that water sufficiently free from metal contaminants can be obtained by passing once-distilled water, rain water, or less satisfactorily, tap water, through columns of synthetic ion-exchange resins (Riches, 1947; Hewitt, 1952).

The mineral salts used by Stout and Arnon were calcium nitrate, potassium nitrate, magnesium sulphate, diammonium phosphate, dipotassium phosphate and ammonium sulphate. Molar solutions of each of these salts were prepared and purified separately, 5 l. at a time, in 6 l. pyrex flasks provided with a plug of cotton-wool. The principle involved in the purification of the solutions was the same as that employed by Steinberg, but the details of the purification varied somewhat for different salts. In all cases 65 g calcium carbonate and a small quantity of a solution of some other salt were added to 5 l. of the solution to be purified. For calcium nitrate and potassium nitrate the solution added along with the calcium carbonate was 50 ml of molar dipotassium phosphate; with dipotassium phosphate and diammonium phosphate the added solution consisted of 25 ml of molar calcium nitrate, and with magnesium sulphate and ammonium sulphate the added solution was 50 ml of molar calcium nitrate + 50 ml of molar dipotassium phosphate. The purification of all the solutions except that of ammonium sulphate was then effected by autoclaving the mixture for an hour at 20 lb pressure, allowing the solution to stand overnight and then filtering. The ammonium sulphate was treated similarly except that the solution was heated for 45 min in a steamer instead of in an autoclave. The final filtrates of the dipotassium and diammonium phosphate were acidified to pH 5·5 with pure sulphuric or nitric acid.

For testing the purity of the solutions so prepared Stout and Arnon made use of dithizone (diphenylthiocarbazone). The testing reagent is prepared by dissolving 0·1 g of purified dithizone in 100 ml of redistilled chloroform. This reagent gives a red or purple colour in the chloroform layer when it is added to a solution containing zinc, copper, lead, nickel, cobalt, cadmium, thallium, mercury or bismuth. By comparing the colour produced by standard solutions with that produced by the purified nutrient solutions, it was found that the latter usually contained less than one part of metal contaminants in 10^8 parts of solution, a degree of purity which was deemed sufficient for culture work on micro-nutrients. Although manganese does not give the colour reaction with dithizone it was concluded that if the other metal contaminants which do produce the colour with dithizone are removed, the manganese will have been removed also.

The salts so purified did not include iron. This was provided as a solution containing 0·5 per cent ferrous sulphate + 0·5 per cent tartaric acid which was added twice weekly to the extent of 0·5 ml per litre of culture solution.

That the method of purification used by Stout and Arnon was justified is clear from the fact that they obtained definite effects which were reproducible in the growth of plants by adding one part of zinc in 2×10^8 parts of culture solution. The method does not, however, provide a satisfactory way for the removal of molybdenum. Hewitt and Jones (1947) at first effected this by precipitation with 8-hydroxyquinoline (oxine) in presence of iron or aluminium but later (1952) used a procedure based on the co-precipitation of molybdenum with copper sulphide. To a molar or twice-molar solution of the major nutrient salts to be purified a 19·5 per cent solution of copper sulphate ($CuSO_4 . 5H_2O$) is added in the proportion of 10 ml to each litre of major nutrient solution. The acidity of the solution is brought to pH 2–4 by addition of hydrochloric acid, and hydrogen sulphide which has been filtered through cotton-wool and then passed through a spray trap is bubbled through the solution for 30 min or for twice this time with calcium salts and phosphates. After precipitation is complete, hydrogen sulphide remaining in the solution is removed by allowing the latter to simmer on a hot plate. This also effects precipitation of colloidal sulphides and sulphur which are removed by filtration. Removal of molybdenum by this means is practically complete and any remaining traces of copper can be removed by extraction with a purified solution of dithizone.

There remains the question of the vessels used to contain the culture solutions. There appears to be a general agreement that containers made of pyrex glass form suitable culture vessels for work on micro-nutrients.

2. The Estimation of Micro-nutrient Elements in Plant Material

In order to investigate the part played by micro-nutrients in plants it is a prerequisite that methods should be available for the quantitative determination of each of them in plant tissues, for without quantitative data little advance in knowledge of any value is likely to accrue. In general, however, the quantities of these elements present in the tissues are so small that the ordinary methods of quantitative chemical analysis are useless for the purpose, and methods have to be found by which very small quantities of the elements concerned can be determined with a reasonable degree of accuracy. Indeed, the advance of plant physiology in general has been retarded very considerably by the lack of methods for measuring many substances in very small quantity, and it is certain that increase of knowledge of the physiology of plants waits in large measure on the development of such micro-methods.

A number of physical instruments have been developed which can be employed by the plant physiologist for the measurement of small quantities of material, and it is now possible with their aid to determine with sufficient accuracy all the known micro-nutrients in plants. These instruments are the absorptiometer, the spectrophotometer, the polarograph and the spectrograph. The absorptiometer is an adaptation of the colorimeter in which the depth of colour of a solution is measured by matching it against that of a standard solution, the matching being made not by the eye, but by a photoelectric cell. The use of this instrument for the determination of small quantities of phosphorus was described by Berenblum and Chain (1938), who showed that quantities as small as $0·1 \mu g$ could be measured with it. The smallest quantities of the trace elements which have so far been determined in this manner are noted later. An authoritative work on the absorptiometer was published by Haywood and Wood in 1944.

With the spectrophotometer the optical density of a solution is measured at a particular wavelength by the combined use of a photometer and a spectrometer. In one form of this instrument parallel beams of light of equal intensity pass through parallel tubes containing the pure solvent and the solution, and then through a photometer. The beams from this are so arranged that they fall on the slit of a spectrometer, so that when viewed through the eyepiece they appear as two spectra one immediately above the other. By means of shutters in the eyepiece any particular part of the spectrum can be isolated for viewing. The photometer is adjusted so that the portions of the two spectra observed are of equal density and the optical density of the solution is then read from a scale on the photometer. The value so obtained is then compared with that of standard solutions. For a description of this type of instrument the work of Strouts, Gilfillan and

Wilson (1955) may be consulted. Brief descriptions of other types of spectrophotometer are given in a work by Jackson (1958).

The polarograph is an instrument in which a solution of an electrolyte in presence of another electrolyte in much higher concentration (known as the ground substance or supporting electrolyte) is subjected to a gradually increasing difference of potential between two electrodes, one of which consists of a series of small drops of mercury delivered from the end of a capillary tube, while the other consists of a still mass of mercury with a comparatively large surface. In these circumstances, when certain experimental conditions are fulfilled a current (the so-called 'wave') flows through the solution when the potential difference reaches a certain value determined by the nature of the cation, or in certain circumstances by the anion, present, while the magnitude of the current is determined by the concentration of these cations (or anions). By means of this instrument it is possible to measure quantities of a number of cations and anions of the order of 1 μg or less. Among the ions which have so far been determined in this way and which are of interest in plant nutritional studies are potassium, sodium, copper, manganese, aluminium, iron, zinc, barium, chloride, sulphate and nitrate, though in general it should be noted that it is not possible with the polarograph to determine one alkali metal in presence of another.

The polarograph was developed by Jaroslav Heyrovsky and appears to have been first described by Heyrovsky and Shikata in 1925. Much of the pioneer work with the instrument was described in English in the *Collection of Czechoslovak Chemical Communications*. Subsequent investigations have shown that this instrument is a most valuable tool for the student of plant nutrition. Several books dealing with methods of polarographic analysis are now available among which those of Kolthoff and Lingane (1952) and Milner (1957) may be mentioned.

The work of Jackson (1958) already cited contains descriptions of a number of methods for the colorimetric determination of manganese, zinc, copper, molybdenum, cobalt, copper and boron, and also for the polarographic determination of zinc and copper, in soils and plant materials. The work of Sandell (1950) on the determination of traces of metals may also be consulted.

Undoubtedly the use of the spectrograph affords the most sensitive method of measuring small quantities of a large number of elements.

Under certain conditions a substance can be made to emit radiation, this radiation being limited to certain wavelengths which are characteristic of the elements in the substance and of the conditions used to excite the radiation. If the radiation passes through a prism or diffraction grating the radiations of different wavelengths are separated and a spectrum results, the radiation possessing the longest wavelengths being at one end of the spectrum and that possessing the shortest wavelengths at the other. In the spectrograph such a spectrum is made to fall on a photographic plate, and the plate, on development, constitutes a photographic negative of the spectrum. This usually consists of a number of lines, each line corresponding to radiation of a definite wavelength and possessing an intensity of blackness depending, within the limits of under-exposure and over-exposure of the plate, on the intensity of radiation of that particular wavelength. Determination of the intensity of blackening of a line characteristic of a particular element should therefore afford a means of estimating the quantity of that element in a sample of material, for example, plant ash, which has been subjected to the necessary conditions for inducing an emission of radiation from it.

Of the various ways in which radiation suitable for spectrographic examination may be produced three have been developed to a considerable extent; these are by means of a flame, by the use of the electric arc and by the use of an electric spark. The spectra produced in these ways are known as flame, arc and spark spectra respectively. Each method has its own particular advantages and disadvantages. For a time at any rate the flame method was perhaps the most popular with workers on plant material, but in more recent years the arc method has been extensively employed for the determination of trace elements in both soils and plant material.

Flame spectra, as the name suggests, are produced when a substance is heated in a flame. Although spectra are produced when the flame is that of a Bunsen burner, for analytical purposes a flame much hotter than this is generally required, and air-acetylene, oxy-coal gas, oxy-hydrogen and oxy-acetylene flames have all been employed with more or less success. Two distinct procedures have been employed for introducing the substance into the flame. In one a small quantity of the substance is contained on or in a piece of filter paper and the latter then introduced into the flame. In the other a solution of the substance is sprayed into the flame through a very fine nozzle. These two methods are largely associated with the names of Ramage and Lundegårdh respectively, but they have both been used, with a variety of modifications in detail, by other workers as well. By maintaining the conditions of experimentation constant the same density of spectral line can be obtained from the same quantity of material, so if calibration is made by the use of a number of samples of known composition it is possible to determine the amount of the element in a sample under examination by measurement of the density of the line and reference to a calibration graph.

In exciting spectra by means of the arc, electrodes in the form of rods, usually of graphite but sometimes of copper, nickel, iron or other metal, and of as great a purity as possible, are used. The electrodes are in a vertical line and the lower contains at its upper end a cavity in which the substance under examination is held. This electrode may be either the anode or cathode, advantages being claimed for each. After bringing the electrodes into contact they are separated to a standard distance apart so that the conditions of the arc are kept as constant as possible. Even so the arc is so variable that it has so far been found impossible to devise an arrangement such that the same amount of material subjected to excitation will produce the same intensity of spectral lines. Hence in quantitative estimation of any element by means of arc spectra it is necessary to have recourse to the device of the 'internal standard'. This is achieved by introducing along with the substance to be examined a known amount of some other substance involving an element which yields a spectral line in the near neighbourhood in the spectrum of the line to be measured. This same consideration holds when the spark discharge is used for exciting spectra. Thus Foster and Horton, in determining boron in plant material by the spark method, added a known quantity of a gold salt to the material they were examining. Within limits the ratio of the intensities of the lines of the element to be measured and of the internal standard is proportional to the ratio of the quantities of the two elements present, so that by measuring the intensities of both lines and reference to a calibration graph obtained by the use of known mixtures in which the quantities of the two elements are varied, the desired determination can be made.

In the buffer procedure as described by Perry, Weddell and Wright (1950) the material to be analysed is mixed with excess of some specific substance and a small quantity of a compound providing the internal standard. This mixture is used in one of the electrodes of the arc. In this way the material subjected to the arc is practically constant, only differing in the quantities of the elements in the soil or plant material, which altogether form only a small proportion of the whole. In using this method Schimp, Connor, Prince and Bear (1957) buffered the plant ash with at least twenty times its weight of chemically pure sodium nitrate and used an alternating current arc. The method was found satisfactory for the determination of a number of elements including manganese and copper. Vanselow and Bradford (1957) mixed 50 mg of the ash sample with 12·5 mg of anhydrous sodium sulphate. They used a direct current arc, the sample to be analysed being usually held in the anode.

Where the content of an element in material to be analysed is too small for it to be determined directly by spectrographic analysis, a method of concentrating it may be employed so that its concentration in the sample analysed is high enough for its estimation. Such methods were divided by Mitchell (1948) into two categories, one which is non-specific in the sense that as many trace elements as possible are concentrated while major constituents are removed, the other in which one or only a few of the minor elements are retained and which may be effected by standard chemical methods of separation. For details of the method developed at the Macaulay Institute for Soil Research reference should be made to the publication cited above.

Mitchell's technique was used by Schimp and his co-workers for the determination of a large number of elements including zinc and molybdenum. It was also used by Vanselow and Bradford for estimations for which their procedure noted above was not sufficiently sensitive. Here again the material to be analysed was mixed with one-third of its weight of sodium sulphate.

For the semi-quantitative analysis of plant ash with the use of the direct current arc Lounamaa (1956) mixed 40 mg of powdered carbon with 20 mg of plant ash and placed the mixture on the anode. For soil analyses a mixture of 4 parts of carbon and one part of sodium chloride was used and this was then mixed with the soil to be analysed in the proportion of one part of soil to 5 parts of the mixture, so that the final mixture placed on the anode contained equal quantities of soil and sodium chloride. The determination of the various trace elements was made by a visual comparison of the densities of the various spectral lines in the photographs of the spectra with those obtained from a series of standards with known contents of the different elements.

For measuring the intensities of the spectral lines various methods have been devised, but this is now generally effected by means of the microphotometer. In this instrument a narrow beam of light passes through the photographic negative of the spectrum, and then falls on a photoelectric cell, with the result that a current is induced which is measured by a galvanometer. By means of a rack and pinion the plate is moved very slowly over the beam of light so that this passes in turn through the clear plate and the spectral line. The difference in the galvanometer deflexion obtained for the clear plate and the spectral line gives a measure of the intensity of the line and hence of the amount of material. The principles of spectrographic analysis have, of course, been given here in the broadest and simplest terms. Actually such analysis is full of difficulties and many precautions have to be taken to ensure reliable results. A description of these details is outside the scope of this book, and those interested should consult works on spectrographic analysis, particularly the publications by F. Twyman (1935, 1938a, 1938b, published by Adam Hilger, Ltd, also 1941), the major works by Lundegårdh (1929, 1934, 1936, 1945), and Lundegårdh and Philipson (1938), and the comprehensive monograph on spectrographic analysis of soils, plants and related materials by Mitchell (1948). This and Lundegårdh's publications are of particular value to workers with biological material. The same remarks apply to the use of the microphotometer and the method of calculating results.

It is to be noted that in the method developed by Lundegårdh solutions are analysed, whereas in most other procedures solid samples are used.

It has already been stated that the spectrograph affords the most sensitive method of measuring small quantities of many elements. This is undoubtedly the case when an arc or spark is used, but results obtained with the flame method indicate that the latter, as used up to now, is capable, broadly speaking, of yielding about the same degree of sensitivity as the polarograph and absorptiometer.

But from Lundegårdh's experience (1929, 1934) it would appear that the flame method is sufficiently sensitive for the quantitative determination of only two of the five generally accepted micro-nutrient elements, namely, manganese and copper. The most dilute solutions of manganese and copper that can be used are those of a concentration of 5×10^{-6} M. As a quantity of 5–15 ml of solution is required this would mean that the smallest quantity of these elements determinable by the flame method as Lundegårdh used it would be of the order of 1·4–4 μg. With a procedure later described by Griggs, Johnstin and Elledge (1941), the minimum concentration of manganese usable is given as $1·125 \times 10^{-5}$ M, while the use of as little as 2 ml of solution is possible. This would mean that the smallest quantity of manganese measurable would be about 1·37 μg.

Using the arc it is possible to measure quantities as small as 0·05 μg or even smaller of manganese, boron, copper, molybdenum, gallium and cobalt. Vanselow and Bradford (1957) found the smallest quantities of the different trace elements detectable by their direct procedure varied considerably, from 0·015 μg for copper to 3 μg for zinc.

It must be borne in mind that for the determination of any particular element one or other of the methods that have been here indicated may be inapplicable. Thus so far it has not been found possible to determine either boron or magnesium, both important plant nutrients, by means of the polarograph. The presence of magnesium as an impurity in graphite may render the use of a graphite arc impracticable for the determination of that element spectrographically. A method, while usable for the determination of larger quantities, may not be sufficiently sensitive for measuring the small amounts of micro-nutrients present in available samples of plant material. Many preliminary trials of the different methods may thus be necessary before the investigator can decide what method to use for the estimation of any particular micro-nutrient.

The degree of accuracy generally obtainable by what may be called micro-methods is much less than that obtainable in most macro-chemical determinations, but is generally sufficient for the kind of problem which faces the investigator of plant nutritional and pathological problems. Broadly speaking it may be said that the results of a single determination obtained by the methods that have been outlined here are correct within about 5–10 per cent. A higher degree of accuracy is no doubt often possible, and with the absorptiometer, polarograph and spectrograph it has been claimed in certain circumstances that the error of a single determination does not exceed 2 per cent. By making a number of replicate determinations and taking the mean value a considerable increase in the accuracy of an estimation may be achieved, but the limited amount of material available may often render this procedure impossible.

A method for the estimation of trace elements in soils and plant materials known as bioassay depends on the effect of these elements on the growth of *Aspergillus niger*. This method has been developed particularly by Mulder in Holland and by workers at Long Ashton in England. Essentially the method consists in growing *A. niger* on culture media which contain all the essential nutrients in adequate quantity with the exception of the one to be determined. A definite quantity of the soil or plant material such as ash is added to the medium and the amount of growth of the fungus after a standard time at a definite temperature is determined from the dry weight of the fungus felt. This is then referred to curves connecting the amount of growth of the fungus with known quantities of the element in question present in the culture medium. This method has been found particularly useful in work with molybdenum, for which it is claimed to be the most sensitive method of determination available. With copper an estimation of the amount of this element can be made simply from visual inspection of the colour of the spores produced, which has been shown to depend on the amount of copper present in the medium. For this work, as with all methods for the determination of trace elements, the utmost obtainable purity of all materials used is essential.

We may next turn to a consideration of the methods suitable for the estimation of the individual micro-nutrients in plant material. They are usually determined in plant ash. This should be prepared by first drying the material in an oven at 70–100 °C, grinding it to a powder and then incinerating it in a furnace at a temperature of from 450 to 600 °C. Vanselow and Bradford (1957) advocated drying the material at 55 °C and then crushing it by hand, which they considered the best way to avoid contamination of the sample. The crushed product was then ignited in a silica dish at a temperature not exceeding 450 °C. Burning in a crucible over a Bunsen burner or blowpipe is not to be recommended, as this leads to an intense local heat causing partial volatilization of some of the mineral matter. The ash, after cooling, is dissolved in a small quantity of mineral acid and the resulting solution evaporated to dryness on a water-bath. The residue is then dissolved in a definite volume of water or dilute acid. Insoluble silica may be removed by filtering.

The preparation of ash in this way is not always to be recommended. For example, Reed and Cummings (1941) found that there was a considerable loss of copper, amounting to 50 per cent or more, involved in this method of ashing, even when the temperature of ignition was as low as 450 °C. For the estimation of copper in plant material they therefore recommended a procedure in which from 0·5 to 2 g of the plant material are heated with 5 ml of concentrated nitric acid until brown fumes are evolved, when 1 ml of concentrated sulphuric acid is added and heating continued until charring begins and all nitric acid is given off. After addition of 1–2 ml of 60 per cent perchloric acid, heating is resumed until the solution is colourless or pale yellow and excess of perchloric acid also given off. The resulting solution should then contain all the copper originally present in the sample used.

It is also possible that such a 'wet ashing' method may be preferable to dry ashing when micro-nutrients other than copper are to be estimated. In this connexion it may be mentioned that Griggs, Johnstin and Elledge (1941) found that dry ashing at 400 °C resulted in a loss of 30 per cent of the total potassium and

10 per cent of the total calcium. They themselves recommended the extraction of the mineral elements with nitric acid and perhydrol. Nitric acid is added to a weighed quantity of the plant material in a 70 ml pyrex test-tube and the mixture heated on a sulphuric acid bath at a temperature between 120 and 140 °C until the solution is clear. Perhydrol is then added and the tube carefully heated over a micro-burner. If necessary, small additions of nitric acid and perhydrol may be made to effect complete decoloration of the solution.

The determination of various trace elements in plant material will now be considered.

Manganese. A considerable number of methods of reasonable accuracy are available for the determination of manganese in small quantities. The spectrograph, polarograph and absorptiometer can all be employed successfully for this purpose. With regard to the spectrograph, the flame, arc and spark methods can all be used, but, as has already been indicated, the sensitivity of the flame method is less than that of the arc. For the determination of manganese in plant material the flame method has been used and described by Lundegårdh (1929, 1934) and more recently by Griggs, Johnstin and Elledge (1941). As already mentioned, by its means about 1·4 μg of manganese can be determined. Lundegårdh claims that the probable error of a single determination made by the flame method is about 1–2 per cent, but the accuracy is no doubt less than this as the amounts determined approach the lower limit of measurable quantities. The line used for the measurement is 4031 Å (Lundegårdh, 1929). Considerably smaller quantities of manganese can be determined by the arc and spark, but although they have been used quite extensively in metallurgical work, they have received relatively little attention from the point of view of the determination of manganese in plants.

Melvin and O'Connor (1941) used the arc method for the simultaneous determination of manganese, boron and copper in fertilizers, and their method would appear to be suitable for the determination of these same trace elements in plant material. The lines used were the manganese 2605·7 Å, boron 2497·7 Å and copper 3247·5 Å, with the beryllium line 3130 Å as internal standard. An accuracy of about ± 5 per cent is claimed for the estimations. Analyses published by the advocates of the method indicate that quantities as small as, or smaller than, 0·1 μg can be determined by means of their procedure. Vanselow and Bradford (1957), who used palladium as an internal standard, considered the limit of detectability of manganese by their method (see p. 26) was 0·03 μg.

Manganese is readily determined with the polarograph, a good, well-defined wave being obtained when a chloride of an alkali or alkaline earth is present in considerable excess. It is usual to employ a few ml of solution, and as concentrations of from $M/250\,000$ to $M/300\,000$ are measurable, with the use of, say, 5 ml of solution 1 μg of manganese is determinable. With the use of special micro-cells taking smaller quantities of solution, very much smaller quantities of manganese can be measured.

But although the polarograph would at first sight appear to offer an ideal way for determining manganese in plants, actually the polarographic determination of manganese in plant ash is not straightforward. This is particularly so where ash or extract contains a considerable quantity of phosphate. Plant ash consists chiefly of oxides, phosphates and sulphates of potassium, calcium, magnesium and other metals and is only soluble in acid. Since hydrogen-ion gives a wave in solutions of alkali or alkaline earth chlorides very near to that of manganese, an ash solution cannot be polarographed for manganese directly because the waves for manganese and hydrogen tend to coalesce. On neutralizing the solution the manganese precipitates as phosphate, and in consequence no wave for manganese is then given. To deal with this situation the following procedure has been found by the writer to give in some cases fairly reasonable results. After removal of sulphate by barium chloride the phosphate is removed from the ash solution by the addition of barium carbonate in excess and filtering. This also removes ferric iron and aluminium, the wave for the latter of which is sufficiently close to that of manganese to render the end-point of the wave for the latter rather indeterminate, even if the two waves do not coalesce. The large quantity of potassium, calcium and magnesium chlorides present in the solution acts as ground substance and the filtered solution can be polarographed directly. Where the quantities of sulphate and phosphate in the ash are relatively small, results obtained by this treatment have been in reasonable agreement with results obtained by other methods, but where much sulphate and phosphate are present results are not reliable, owing probably to adsorption of manganese by barium sulphate or phosphate. For each determination two solutions are taken, one consisting of ash solution, the other consisting of ash solution containing the same concentration of ash but with a known amount of added manganese. The difference in the heights of wave given by the two solutions is then attributable to the added manganese, to which the height of the wave given by the pure ash solution is referred.

Since a number of metals are reduced at more positive potentials than manganese there is the possibility that they might interfere with the wave for this ion. The principal elements concerned are copper, cadmium, lead, chromium, molybdenum, cobalt, nickel, iron and zinc. It is extremely unlikely that any of these, with the exception of the last two, are likely to occur in sufficient quantity in plant material to disturb the polarographic determination of manganese. As regards iron this is practically all removed by the treatment of the ash solution outlined above, and as a matter of fact no wave for iron appears in the polarograms of solutions so treated. The possible influence of zinc on the manganese wave was examined by the writer, who found that no effect was produced on the wave of $M/10000$ manganese by zinc in concentrations up to five times that amount. No ash examined by the writer has been found to contain a proportion of zinc to manganese approaching that value. According to Jackson (1958) the manganese content of plant digests can be determined satisfactorily with the polarograph with the use of a ground solution containing $0.25M$ sodium sulphite and $0.1M$ sodium chloride. It is claimed that manganese in concentrations from 0.01 to 0.1 millimols per litre can then be measured with an accuracy of about 5 per cent in presence of copper and zinc in similar concentrations.

A number of methods for determining manganese with the use of the absorptiometer are available. Most of these depend on the oxidation of the manganese salt with the production of permanganate, the intensity of the colour of which is determined with the absorptiometer. The methods are known as the periodate, persulphate, and bismuthate methods, according to the reagent used for effecting the oxidation. Trials carried out by my former colleagues, Dr K. W. Dent and Professor E. S. Twyman, indicate that the periodate method is the most satisfactory of these, as regards both simplicity of procedure and accuracy of the results obtained. This experience appears to be fairly general if we are to judge from the fact that this method is the one most commonly used by recent workers. A number of accounts of the procedure employed in using the method have been published; of these may be cited those by Coleman and Gilbert (1939) and Cook (1941). Coleman and Gilbert found that the same values for manganese content of tea and coffee were obtained with a wet ashing process and with incineration in a muffle, so it is concluded that no loss of manganese occurs with either ashing process. When dry ashing is employed 1 g of material is moistened with 1–2 ml of concentrated sulphuric acid and ashed at a dull red heat in a muffle. The ash is dissolved in 15 per cent sulphuric acid and filtered. The filter paper is ignited and any residue from this dissolved in about 10 ml of dilute sulphuric acid and filtered into the main filtrate. To the combined filtrate 0·1 g of potassium iodate is added and the mixture boiled for a few minutes and then kept hot for 30 min for the development of the pink colour. The solution is made up to standard volume and the intensity of the colour determined.

Amounts of manganese down to about 12 μg can be estimated in this way.

For a description of the bismuthate method, which is particularly recommended for the determination of manganese in soils, reference may be made to a paper by Dean and Truog (1935) and for the persulphate method to papers by Majdel (1930) and Olsen (1934).

A method for the estimation of small quantities of manganese which has been described by Sideris (1940) depends on the colour produced when formaldoxime is added to a solution of a manganese salt. The formaldoxime reagent is prepared by dissolving (by boiling) 20 g of trioxymethylene + 47 g of hydroxylamine sulphate in 100 ml of distilled water. Ten ml of the hydrochloric acid solution of the ash are neutralized with sodium hydroxide and then acidified with 2 ml of a 20 per cent solution of acetic acid. Excess phosphate is then removed by adding 0·5 ml of a 5 per cent solution of lead acetate, the mixture shaken, allowed to stand for 10 min and then treated with 1 ml of a 20 per cent solution of sodium sulphate to remove excess lead. After 30 min the precipitate is removed by filtration or centrifuging and the clear solution neutralized with 40 per cent sodium hydroxide. Three or four drops of formaldoxime reagent are added to the liquid and then more 40 per cent sodium hydroxide until a wine-red colour develops, the intensity of which is said to be directly proportional to the concentration of manganese. The liquid is made up to a standard volume and the intensity of the colour determined. According to Sideris it would appear that quantities of manganese of the order of 5 μg can be determined in the presence of 10–100 μg of phosphate with an error not exceeding 4 per cent. Under more favourable conditions quantities of manganese down to 0·25 μg would appear to be determinable.

Yet another method for the determination of small amounts of manganese has been described by Wiese and Johnson (1939). The nitric acid solution of the ash containing from 1 to 10 μg of manganese is first rendered free from chlorides by three times evaporating it to dryness and redissolving in nitric acid + 10 ml of distilled water. About 0·2 g of sodium bismuthate is added and the mixture boiled for 2–3 min. On cooling to below 30 °C 0·2–0·3 g of sodium bismuthate is added, and after mixing the sample thoroughly and allowing it to stand for a few minutes the excess of sodium bismuthate is filtered off through a Gooch crucible. The solution filters directly into the absorptiometer cell containing two drops of a solution of benzidine (1 per cent in 5 per cent acetic acid) in 3 ml of distilled water. The solution is made up to standard volume and the intensity of the yellow-green colour which develops is estimated after 5 min.

Zinc. For the determination of zinc in plant material by means of the spectrograph, the flame method is not sufficiently sensitive. Several workers, however, have described procedures for the spectrographic determination of zinc in such material by using the arc. Thus Rogers (1935) advocated using the zinc line 2138·5 Å but found it was necessary to sensitize the photographic plate by spreading mineral oil over the emulsion, or alternatively using a special plate (Eastman spectroscopic plate type III-0) with ultra-violet sensitization. The tellurium line 2143·0 Å was used as internal standard. Vanselow and Laurance (1936), on the other hand, recommended the use of the zinc line 3345·0 Å with cadmium line 3252·5 Å as internal standard. To a hydrochloric acid solution of plant ash a known amount of cadmium sulphate was added and the zinc and cadmium then precipitated as sulphides by a special technique; the sulphides were then spectrographed. Rogers and Gall (1937) reported unfavourably on the procedure of Vanselow and Laurance and suggested that zinc in plant ash is not completely extracted by hydrochloric acid. O'Connor (1941), in developing spectrochemical methods for the determination of trace elements in fertilizers, like Rogers used the zinc line 2138·5 Å for the determination of this element, but employed the beryllium line 2348·6 Å as internal standard. The spectrograms were obtained on photographic plates with ultra-violet sensitization (Eastman I-0 spectroscopic ultra-violet sensitive plate). O'Connor claims that in this way zinc can be determined within the limits of 2 p.p.m. to 1 per cent with an accuracy within ± 5 per cent. As a 20 mg sample is used for a determination this means that a quantity of zinc as small as 0·04 μg can be determined. Later Vanselow and Bradford (1957) recommended the thallium lines 2767·9 Å and 3519·2 Å as internal standards, while Schimp, Connor, Prince and Bear (1957) used the germanium line 3269·5 Å for this purpose.

Lundegårdh (1934) recommended the use of the line 3345·0 Å for the determination of zinc by the spark method.

Methods for the polarographic determination of zinc in plant materials have been elaborated by Stout, Levy and Williams (1938), by Reed and Cummings (1940) and by Walkley (1942).

In the method of Stout, Levy and Williams the hydrochloric acid ash solution (about 100 ml) from 1 to 2 g of dried plant material is treated with 5 ml of N ammonium citrate and then rendered slightly alkaline by the addition of ammonium hydroxide. The resulting solution is then shaken with 10 ml of a solution of 1–3 mg of dithizone in chloroform. The resulting chloroform layer, which then contains the zinc, nickel, cadmium, lead and copper, is separated from the aqueous layer containing iron and manganese. The zinc and accompanying metals are then removed from the chloroform by two extractions with

10 ml of 0·5 N hydrochloric acid. The hydrochloric acid extracts are evaporated to dryness and the residue dissolved in a solution of 0·1 N ammonium acetate + 0·025 N potassium thiocyanate. On polarographing this solution well-defined and well-separated waves of lead, cadmium, nickel and zinc are obtained.

Reed and Cummings experienced some trouble with the method of Stout, Levy and Williams, particularly when large quantities of zinc or of cadmium, lead, copper, nickel or cobalt were present, and they found that to effect a quantitative separation of the zinc from aluminium, iron and alkali metals five or six extractions, instead of only one, were necessary. They therefore devised a different procedure in which the ash solution in hydrochloric acid was brought to a pH of between 4 and 5 by addition of dilute ammonium hydroxide. This precipitates practically all the aluminium and ferric iron, which are filtered off. The filtrate is evaporated to dryness on a steam bath and the residue dissolved in a solution of 0·1 N ammonium acetate of pH 4·6 and containing also potassium thiocyanate of concentration 0·025 N, and the resulting solution polarographed. It is stated that no interference with the height of the zinc wave then results from the presence of chloride, sulphate, phosphate, carbonate, sodium, potassium, calcium, magnesium or manganese, and no interference results from lead, cadmium or nickel up to concentrations ten times that of the zinc. Copper interferes if it is present in concentrations ten times or more that of zinc, while the cobalt and zinc waves tend to coalesce if the former is present in relatively high concentration. Actually Reed and Cummings found that if cobalt occurs in concentrations greater than 1×10^{-5} g per ml it only interferes with determinations of zinc when the ratio of cobalt to zinc exceeds 2. It would, as a matter of fact, be a very exceptional plant ash in which any of the ions noted occurred in sufficient amount to interfere with the polarographic determination of zinc in this way.

The lowest measurable concentration of zinc is stated by Reed and Cummings to be 0·2 μg of zinc per ml, that is, about M/300000, so that, using 5 ml of solution, 1 μg of zinc should be determinable.

In Walkley's procedure the dried plant material is first subjected to wet ashing, about a gram of the material being digested with a mixture of 10 ml of nitric acid, 1 ml of sulphuric acid and 1 ml of perchloric acid. Frothing is prevented by addition of a drop of kerosene. The cooled digest is taken up in 15 ml of water and boiled, and on cooling 25 ml of an ammonium citrate buffer are added. This buffer is prepared by dissolving 5 g of citric acid in 50 ml of water and 200 ml of 4N ammonium hydroxide, and then extracting impurities by three successive shakings with 10 ml of a solution of dithizone in chloroform (1 g of dithizone in 100 ml of chloroform), and running off the chloroform layers. The final purified buffer solution contains dithizone in solution.

The zinc is separated from the digest after treatment with the ammonium citrate buffer by three extractions each with 5 ml of chloroform. The chloroform extracts are evaporated to dryness, then treated with a mixture of 2·5 ml of nitric acid, 0·5 ml of perchloric acid, and two drops of sulphuric acid and again evaporated to dryness, boiling being avoided. The residue is then dissolved in 1 ml of a ground liquid of 0·1N ammonium chloride + 0·02N potassium thiocyanate containing 0·0002 per cent of methyl red and polarographed in a small electrolysis vessel. Practical details for carrying out Walkley's method will be found in Piper's book on *Soil and Plant Analysis* (1942 *b*).

For determining zinc and copper on a single polarogram Menzel and Jackson (1951) also used a wet ashing process in which a preliminary digestion of the dried plant material with concentrated nitric acid at 100 °C was followed by further digestion at the same temperature with a mixture of the same three acids used by previous workers, but this time concentrated nitric acid, concentrated sulphuric acid and 60 per cent perchloric acid were used in the proportions 10:1:4. The digest from 1 or 2 g of dried material was brought into a separating funnel with 25 ml of ammonium citrate buffer containing dithizone and 5 ml of chloroform, precautions having being taken to ensure the purity of the reagents. Most of the zinc and copper salts with dithizone are extracted by the chloroform with shaking for a minute and the pH of the aqueous phase brought to 9–10 by addition of either hydrochloric acid or ammonium hydrate. After a further shaking for a minute the chloroform layer is removed and the aqueous layer washed with 2 ml of chloroform and then with a further 2 ml of this, by which all the zinc and copper will have been transferred to the chloroform. The chloroform solution of the zinc and copper dithizone compounds is evaporated to dryness and the dithizone oxidized by heating with the three-acid mixture noted above at 300 °C for 2 h. For polarographing, the dry sample so obtained is added with a drop of gelatin to 5 ml of the ground liquid which is prepared by dissolving 2·1 g of sodium sulphite in 66 ml of 0·1M ammonium hydroxide. About an hour, with occasional gentle shaking of the liquid, may be necessary to bring the zinc and copper into solution. Where the copper content of the sample is low a smaller amount of the ground liquid may be used in order to raise the concentration of the copper in the solution polarographed.

For the absorptiometric determination of zinc a favoured method is that based on the coloration given by zinc salts with dithizone (diphenylthiocarbazone). Reference has already been made in this chapter to the fact that this reagent gives a red or purple colour when added to solutions of compounds of a number of metals, including zinc (see p. 20). The resulting coloured compound is quantitatively extracted with chloroform or carbon tetrachloride, and according to R. H. Caughley (see Holland and Ritchie, 1939) sodium diethyldithiocarbamate in 0·02N ammonium hydroxide solution inhibits the reaction of dithizone with all metals except zinc. Cowling and Miller (1941) made use of this very useful fact to work out an absorptiometric method for the quantitative estimation of zinc in plant materials. Unfortunately the addition of the sodium diethyldithiocarbamate (generally denoted by 'carbamate' for the sake of brevity) renders the extraction of zinc by dithizone incomplete. By carefully standardizing the technique, however, Cowling and Miller claimed that this drawback could be overcome and the quantity of zinc in plant ash determined with reasonable accuracy.

After ashing a 5 g sample of the dried plant material at 500–550 °C, the ash is dissolved in hydrochloric acid, insoluble matter being removed by filtration. The metals which form complexes with dithizone are then extracted from the solution by repeated treatment with excess of a solution of dithizone in carbon tetrachloride at a pH of 8·5–9 in presence of ammonium citrate; the latter prevents the precipitation of iron and aluminium. The dithizone extract is then treated with 0·02N hydrochloric acid;

copper and the excess of dithizone remain in the carbon tetra-chloride phase while the zinc and other metals pass into the aqueous phase. The pH of the latter is then adjusted to between 8·5 and 9 by the addition of ammonia-ammonium citrate buffer containing carbamate and the zinc extracted with dithizone in carbon tetrachloride. The resulting extract is used for the determination of the zinc, the reading obtained being compared with those given by solutions of known zinc content which are plotted to give a standard curve. It is essential that the same conditions should be rigidly adhered to both in obtaining the standard curve and in the analysis of samples of plant material; that is, the same pH should be used in the extraction, the volumes of phases, the amount of dithizone and the amount of carbamate used should be the same. Tests made by the authors show that if this is done the method is highly reliable. The presence of other metals does not interfere significantly with the determination; a good degree of accuracy was obtained in the recovery of zinc added to plant material, while good agreement was obtained between determinations of zinc in duplicate samples of the same material. The method would appear to be capable of determining quantities of zinc as small as 2 μg or even less.

A method which can be used for the colorimetric determination of both zinc and copper, involving the use of 2-carboxy-2'-hydroxy-5' sulphoformazylbenzene, conveniently known as zincon, has been described by Rush and Yoe (1954). With this reagent both zinc and copper form a blue complex, that with zinc being stable over the pH range 8·5 to 9·5 while the complex with copper is stable over the wider range of 5·0 to 9·5. For the determination of zinc and copper in a solution containing both these elements 10 ml of a Clark and Lubs buffer at pH 5·2 and 3 ml of a zincon solution are added to 10 ml of an approximately neutral solution, which should contain from 25 to 100 μg of zinc and copper altogether, and the mixture is then made up to a standard volume of 50 ml. The depth of the colour is then measured with a spectrophotometer at 600 mμ. This gives a value for the copper. The operation is repeated with another 10 ml of the solution but this time it is buffered at pH 9·0 at which both the copper and zinc complexes are stable so that a value is obtained for copper + zinc. The value for zinc is thus obtained by difference. The method was used by Bingham, Martin and Chastain (1958) for the estimation of these elements in *Citrus* leaves.

Copper. The spectrographic determination of copper in plant material can be carried out by both the flame and arc methods. With the use of the Lundegårdh flame method the sensitivity, according to the experience of Griggs, Johnstin and Elledge (1941), is about half that found for manganese, these workers giving the minimum concentration of copper usable as 0.000025 M. As the minimum quantity of solution which can be employed in their arrangement is 2 ml this would give the lowest measurable amount of copper to be 3.15 μg. The line used for the measurements is 3247.5 Å (Lundegårdh, 1929).

It has already been mentioned (p. 31) that with the use of the arc Melvin and O'Connor (1941) have achieved the simultaneous determination of boron, manganese and copper. The copper line used for the determinations was 3248 Å, and the beryllium line 3130 Å was used to provide the internal standard. From the analyses published by Melvin and O'Connor it would appear that the copper forming 0.001 per cent of a 10 mg sample of material can be measured, from which it would seem that quantities of copper of the order of 0.1 μg can be estimated. As mentioned earlier Vanselow and Bradford (1957) estimated that the smallest quantity of copper detectable by their direct method was 0.015 μg. Palladium and germanium provided their internal standards.

For the spectrographic determination of copper in grasses Rusoff, Rogers and Gaddum (1937), employing the arc, used cadmium as internal standard.

A procedure for the estimation of copper in plant materials by means of the polarograph has been described by Reed and Cummings (1941). The solution obtained by wet ashing of from 0.5 to 2 g of plant material (see p. 30) is diluted to 15–20 ml, heated to boiling and then rendered alkaline by the addition of a slight excess of concentrated ammonium hydroxide. The solution is then boiled for a minute and filtered, the residue being washed with slightly ammoniacal water and the washings added to the filtrate. The latter is evaporated to dryness and the resulting residue dissolved in 10 ml of a ground liquid consisting of 4.5 ml of 0.5 M sodium hydroxide + 4.5 ml of 0.5 M citric acid + 1 ml of 0·05 per cent of acid fuchsin. This liquid, when polarographed in absence of oxygen, gives a well-defined wave for copper. In this way, according to Reed and Cummings, concentrations of copper down to about 3×10^{-6}M can be determined, corresponding to about 2 p.p.m. of copper in the plant. Published data indicate that in normal plants the amount of copper is usually well above this value, but in plants showing symptoms of copper deficiency it may be lower than this. In such cases it would be necessary either to use larger quantities of plant material for the determination or to use a method other than a polarographic one. The procedure for the simultaneous determination of copper and zinc by means of the polarograph proposed by Menzel and Jackson has already been indicated.

The absorptiometric method generally used for the determination of copper in plant material depends on the intense colour produced by copper salts with diethyldithiocarbamate (Callan and Henderson, 1929), the coloured complex being extracted with amyl alcohol. Procedures particularly applicable to biological material have been described by Eisler, Rosdahl and Theorell (1936), by Eden and Green (1940), and by Piper (1942 a, b). In the procedure of the first group of workers the dried material is heated on a sand-bath with sulphuric acid and perchloric acid until the mixture is colourless or pale yellow. The resulting liquid is cooled under the tap and then rendered slightly alkaline to litmus by the addition of 8–9 per cent ammonium hydroxide. Iron is precipitated by treating the alkaline solution with a few ml of a 4 per cent solution of sodium pyrophosphate, then heating at 80 °C for 30 min and cooling to room temperature. If a crystalline precipitate is present at this stage it must be dissolved by the addition of water. Now 2 ml of water and 5 ml of amyl alcohol are added and then immediately 0·5 ml of a 2 per cent solution of sodium diethyldithiocarbamate.

The mixture is now strongly shaken and centrifuged. The amyl alcohol layer is separated and its light absorption in light of wavelength 4400 Å measured. The depth of colour of the amyl alcohol layer is dependent not only on the concentration of copper, but also on the salt concentration of the aqueous phase from which the amyl alcohol layer is separated, the salt concentration depending on the amount of sulphuric acid used. Hence it is necessary to use a constant technique in order to obtain reliable results.

Eden and Green also used a wet ashing method, the material, for example 5 ml of blood, 1–5 g of fresh tissue or 1 g of dried material, being digested with a mixture of sulphuric and perchloric acids to which nitric acid is added later. After dilution with water the digest is treated with 2 ml of 50 per cent ammonium citrate and 5 ml of ammonia (sp. gr. 0·880). The ammonium citrate effects the deionization of the iron and prevents the precipitation of calcium phosphate. The solution is then made up to 25 ml with water. If the material is relatively low in calcium and phosphorus but high in iron it is preferable to use 10 ml of 4 per cent hydrated sodium pyrophosphate instead of the ammonium citrate, as the resulting solution is colourless, whereas with citrate the solution is slightly coloured.

Two ml of a recently filtered 0·5 per cent solution of sodium diethyldithiocarbamate are now added to the solution, kept constantly shaken, then 5 ml of amyl alcohol added, and the mixture vigorously shaken for 30 sec. The amyl alcohol extract, which contains the coloured copper diethyldithiocarbamate, is centrifuged or filtered through acid-extracted filter paper to remove any water in suspension and the depth of colour determined with a colorimeter or absorptiometer by comparison with a standard solution of copper diethyldithiocarbamate prepared in the same way as that from the tissue digests. With adequate precautions it would appear that quantities of copper as small as 0·3 μg are determinable by the procedure of Eden and Green.

In the method as used by Piper a wet ashing process is also used, and after diluting the digest with water and treating it with ammonium citrate solution to dissolve any hydrolysed manganese and to prevent precipitation of phosphates, the resulting solution is brought to pH 3 and the copper extracted with dithizone in carbon tetrachloride. The carbon tetrachloride is then removed by evaporation and the dithizone by heating with a little sulphuric acid containing a drop or two of perchloric acid. After dilution with water and rendering the solution alkaline with ammonium hydroxide, the copper is precipitated by the addition of a few drops of a 3 per cent aqueous solution of sodium diethyldithiocarbamate. The copper diethyldithiocarbamate is then extracted with amyl alcohol and the intensity of colour of the extract compared with that of a standard. For full details of the experimental procedure reference should be made to Piper's book. The procedure involving the formation of a coloured complex of copper with zincon has already been described (p. 40).

It has been mentioned earlier that in the bioassay of copper with the use of *Aspergillus niger* as test organism, an estimate of the amount of copper could be made from the colour of the spores. The relation between the copper concentration of the medium and spore colour observed by Nicholas and Fielding (1950) is indicated in Table 6.

TABLE 6. *Effect of copper concentration in the medium on the colour of the spores of* Aspergillus niger.
(From Nicholas and Fielding)

Copper concentration as μg in 40 ml of medium	Colour of spores
0	No spores
0·25	Orange-yellow
0·5	Light brown
0·75	Greyish brown
1·0	Dark brown
2·5	Black
5·0	Black

Boron. As regards the determination of boron by means of the spectrograph, the usual flame method is useless. Even with a 0·1M solution of boric acid no boron line was obtained by Lundegårdh in a flame spectrum. Boron in plant material has, however, been determined by means of the spark method by Foster and Horton (1937). The plant material was used fresh, being neither dried nor ashed, but crushed and disintegrated into a fine pulp in a small copper mortar. The samples used consisted of 100 mg of fresh material. Measurements were made on the boron line 2497·7 Å, and the gold line 2427·9 Å was used as internal standard. Six replicate determinations of the boron in turnip leaves gave values varying from 3·7 to 4·3 μg of boron per g of fresh tissue, with an average value of 4·1 μg. Foster and Horton concluded that the error of a single determination was not in excess of 10 per cent. The limit of sensitivity of the method was not clearly stated, but it would seem probable that quantities of boron as small as 0·1 μg might be measurable in this way. Lundegårdh (1929) suggested the use of the cadmium line 2573 Å as internal standard in the determination of boron by the spark method.

As mentioned earlier (see p. 31), Melvin and O'Connor (1941) have used the arc method for the simultaneous determination of boron, manganese and copper in fertilizers, using beryllium as internal standard. The method would seem to be applicable to the determination of boron in plant material, and the accuracy and sensitivity would appear to be of the same order as in the spark method of Foster and Horton.

Although the flame method is not suitable for the spectrographic determination of boron, McHargue and Calfee (1932) have successfully made use of flame spectra for the optical spectroscopic determination of boron. In their earlier procedure the boron was first converted into methyl borate, then volatilized with methyl alcohol and burnt in an atmosphere of oxygen in front of a cell containing a solution of potassium permanganate. The concentration of this solution was adjusted so that the bright lines of the boron spectrum were just obscured by it. By previous standardization of solutions of potassium permanganate against standard boron solutions the concentration of the experimental boron solution was obtained.

Later (Calfee and McHargue, 1937) a different procedure was devised. The spectrum was excited in an oxygen-methane flame, methane saturated with a solution of methyl borate in methyl alcohol being ignited in an oxygen blast. The light emitted from this was polarized and by an optical arrangement the spectrum was brought into juxtaposition with a second spectrum similarly produced by the burning of a standard boron solution. By rotation of an analysing plate the intensity of the spectrum of the standard boron solution could be varied, and so matched with that of the solution the boron concentration of which it was desired to measure. For a quantitative determination in this way the boron content of a sample should lie between 25 and 50 μg. Agreement between the two procedures was good, but the second method was found to be more exact.

So far it has not been found possible to estimate boron polarographically.

A colorimetric method for the determination of small quantities of boron which has been applied to the estimation of boron in plant material depends on the colour change of quinalizarin effected by boric acid (see, for example, Smith, 1935). To 1 ml of solution containing from 1 to 40 μg of boric acid 9 ml of concentrated sulphuric acid are added followed by 0·5 ml of a 0·01 per cent solution of quinalizarin in 93 per cent sulphuric acid. A colour change from reddish violet to blue results, the complete process taking about 5 min. Nitrate, dichromate and fluoride must not be present, but the common metals do not interfere with the reaction. (See also Berger and Truog, 1939, 1944).

While this method appears to be quite satisfactory for the estimation of boron in plant material, the necessity of using concentrated sulphuric acid is something of a drawback. Another colorimetric method for the determination of small quantities of boron, which does not involve the use of this reagent, was described by Naftel (1939). This method depends on the colour produced when a solution of boric acid is treated with oxalic acid and either curcumin or an extract of turmeric and the mixture evaporated to dryness. According to the procedure recommended by Naftel, the soil or plant-ash extract containing from 0·5 to 8 μg of boron is first rendered alkaline by the addition of 5 ml or more of 0·1 N calcium hydroxide and then evaporated to dryness on a water-bath. After cooling there are added to the residue 1 ml of a freshly prepared solution of oxalic and hydrochloric acids (made by adding 80 ml of a saturated solution of oxalic acid to 20 ml of concentrated hydrochloric acid) and 2 ml of a 0·1 per cent solution of curcumin or a 1 per cent freshly prepared extract of turmeric in 95 per cent ethyl alcohol. The mixture is evaporated to dryness on a water-bath at 55 °C, heated for a further 30 min at this temperature and then cooled, extracted with 95 per cent ethyl alcohol and the colour of the clear solution obtained after filtering or centrifuging compared with that of standard solutions prepared in the same manner. Quantities of boron down to 0·5 μg can be determined by this method. In soils other elements present do not appear to interfere with the estimation of boron, but if large quantities of other substances present are found to interfere with the determination of boron, the latter may first be separated by volatilization with methyl alcohol. A procedure involving this method has been described more recently by Dible, Truog and Berger (1954).

A method described by Hatcher and Wilcox (1950) depends on the colour change produced in a solution of carmine in sulphuric acid by the presence of boron, a stable purple or blue resulting. The solution to be analysed should contain not more than 20 μg of boron and should be diluted or concentrated if necessary in order to bring the boron concentration within the desired range. The depth of the colour resulting is measured by a spectrophotometer at 585 mμ.

An electrometric titration method has been specially devised by Wilcox (1940) for the determination of boron in plant material. A quantity of the dried and powdered plant material (say, from 5 to 25 g) containing not more than 2 mg of boron is mixed with one-tenth of its weight of calcium oxide and ignited in a furnace at a low red heat. The resulting ash after cooling is moistened with water and taken up in 15–20 ml of 6 N hydrochloric acid and then heated on a steam-bath for 30 min. Phosphate is removed from the resulting solution by the addition of N lead nitrate solution to the extent of 1 ml for each gram of plant material used, followed by sodium bicarbonate until a precipitate is produced, when the mixture is heated on a steam-bath and more sodium bicarbonate added until the solution is neutral to brom-thymol-blue (about pH 7). The mixture is then made up to 250 ml and filtered through a dry filter paper. Carbon dioxide is removed by acidifying with 6 N hydrochloric acid and heating to boiling, then making alkaline with 0·5 N sodium hydroxide and reacidifying with 2 N hydrochloric acid until 5–10 drops in excess have been added. On making up to 300 ml the solution is boiled for a few minutes. It is then ready for electrometric titration. For this the quinhydrone electrode may be used in conjunction with a 0·7 N calomel electrode, these giving a null point at approximately pH 7. The electrodes having been introduced into the solution, 0·5 N sodium hydroxide is added until the solution is neutral to brom-thymol-blue, when the galvanometer should register approximately zero; if it does not the null point is obtained by the addition to the solution of either 0·0231 N sodium hydroxide or hydrochloric acid. Five grams of mannitol are then added, and if boric acid is present a galvanometer deflexion results. Standard

0·0231 N sodium hydroxide is then added until the null point is again reached. The volume of sodium hydroxide gives a measure of the amount of boron present, 1 ml 0·0231 N sodium hydroxide being equivalent to 0·25 mg boron. This method is claimed by Wilcox to be specially suitable for determination of boron in tissues where this element is present in quantity less than 50 p.p.m.

Molybdenum. In determining molybdenum spectrographiccally by means of the arc Vanselow and Bradford, using the lines 3132·6 Å, 3170·3 Å and 3208 Å with germanium and palladium providing internal standards reported the limit of detectability as 0·06 μg.

With the use of the colorimeter or absorptiometer molybdenum in plant material has generally been estimated as molybdenum thiocyanate by the method described by Marmoy (1939). A 50 ml sample of the hydrochloric acid solution of the ash containing not more than 20 μg of molybdenum and having an acid concentration of 14 per cent by volume is treated first with 3 ml of potassium thiocyanate and then with 3 ml of stannous chloride. The molybdenum thiocyanate produced is then extracted with ether, the extractions being repeated until the ether layer is colourless, and the depth of colour of the combined ether extract is compared with that of a standard solution in ether of molybdenum thiocyanate prepared from ammonium molybdate in the same way as that from the ash. A modification of the method in which ammonium thiocyanate is used has been described and employed by Purvis and Peterson (1956) for the estimation of molybdenum in soils and plant material. Concentrations of molybdenum in lucerne ranging from 0·23 to 1·93 p.p.m. were determined in this way.

Another colorimetric method for determining small quantities of molybdenum was described by Yoe and Will (1952). This depended on the fact that a substance disodium-1,2-dihydroxy-3,5-benzenedisulphate, conveniently designated as tiron, gives a bright yellow colour with molybdenum ions. It was claimed that under suitable conditions molybdenum could be measured down to a concentration of 0·1 p.p.m.

As mentioned earlier, the method of bioassay has been found particularly useful for the determination of molybdenum. Details of this method and examples of its use have been published by Mulder (1948), Nicholas and Fielding (1950) and Hewitt and Hallas (1951). The last-named workers claimed that in this way it was possible consistently to detect 0·000 15 μg of molybdenum and by interpolation to estimate molybdenum to the nearest 0·000 05 μg.

Aluminium. Aluminium is present in quantity in many soils and is generally present in plants, but so far its essentiality has been indicated for only a few species. Perhaps for this reason not so much attention has been given to the determination of aluminium in plant material during recent years as to the better established micro-nutrients.

For the measurement of small quantities of aluminium, such as might be expected in samples of plant material, direct measurement of the intensity of the aluminium line in the air-acetylene flame spectrum is not suitable, for Lundegårdh (1929) found that no line was obtained with even a 0·1M solution of aluminium chloride. Mitchell and Robertson (1936) have, however, described a means by which aluminium in concentrations ranging from about 2 to 10 mg per litre can be determined by the Lundegårdh method. It depends on the fact that the presence of aluminium brings about a lessening of the intensity of the calcium and strontium lines of the flame spectrum, the decrease in intensity varying with the amount of aluminium present and also with the relative amounts of calcium and strontium present. Hence with careful control of the conditions the depression in the intensity of these lines can be used to determine the aluminium content of solutions of the concentrations indicated above. Using 15 ml of solution for a determination, this means that quantities down to 30 μg of aluminium can be measured in this way.

With the spark the best aluminium line for measurement is, according to Lundegårdh (1934), 3961·5 Å, but if the sample under examination has a high calcium content, which may frequently be so with plant material, the calcium line 3968·5 Å may interfere with the aluminium line.

Aluminium may be determined polarographically with the use of lithium chloride, barium chloride or magnesium chloride as supporting electrolyte, but, owing to the fact that the aluminium wave occurs at a rather high negative potential as well as to difficulties resulting from the presence of phosphates, it is unlikely that the polarograph will afford a simple means for the determination of aluminium in plant material.

Two colorimetric methods, suitable for use with the absorptiometer, depend on the formation of lakes, fairly stable in the presence of acetic acid, when alizarin and the ammonium salt of aurin tricarboxylic acid, respectively, are added to a solution of an aluminium salt. Both appear to be adaptable to the determination of aluminium in plant tissues. The first method appears to have been described first by Atack (1915). The modification of it described by Underhill and Peterman (1929) was used in my laboratory on fairly pure solutions with marked success. The second method was described by Hammett and Sottery (1925) and was adapted for the determination of aluminium in animal tissues by Myers, Mull and Morrison (1928). Amounts of aluminium of the order of 5 μg can be measured by both methods.

A colorimetric method depending on the use of eriochrome cyanine R has been applied by Jones and Thurman (1957) to the estimation of aluminium in soil and plant materials. For this are required an iron-compensating solution, a solution of sodium thioglycollate as well as a solution of the eriochrome cyanine R. The iron-compensating solution is prepared by dissolving 0·7022 g of ferrous ammonium sulphate in about 100 ml of distilled water and 5 ml of concentrated sulphuric acid contained in a silica basin, and the ferrous iron is then oxidized to ferric iron by the addition of 5 ml of concentrated nitric acid. After warming the solution to remove oxides of nitrogen the liquid is cooled and made up to 1 litre. The sodium thioglycollate solution is prepared by dissolving 2·5 g of this substance in 200 ml of water, adding 125 ml of 95 per cent ethanol and making the solution up to 500 ml. The eriochrome cyanine R solution is obtained by dissolving 0·75 g of the substance in about 200 ml of distilled water, adding 25 g each of sodium chloride and ammonium nitrate and then 2 ml of concentrated nitric acid. The solution is then made up to 1 litre.

For the determination of aluminium, to 5 or 10 ml of a solution prepared from soil or plant material there are added 2 ml of the iron-compensating solution, 10 ml of the sodium thioglycollate solution and 5 ml of the eriochrome cyanine R solution. The solution is thoroughly mixed and brought to pH 6 by the addition of 10 ml of ammonium acetate buffer. After 18 min the absorption is measured at 535 mμ with a spectrophotometer and the concentration of the aluminium determined by reference to a calibration curve obtained by the use of standard aluminium solutions. The maximum intensity of the colour developed is reached in 10 min after bringing the aluminium solution and the eriochrome cyanine R together and remains so for another 30 min. It is claimed that the method is accurate to 0·5 μg.

Cobalt and nickel. Although evidence has occasionally been adduced to indicate that small quantities of cobalt and nickel may bring about an increase in the rate of growth of plants, there has up to now been no definite proof that either of these elements is essential for the growth of any higher plant. There is, however, very definite evidence that cobalt is essential for sheep and cattle, and as the deficiency of this element in the animal must arise from the low content of cobalt in the plants on which the animal feeds, the determination of small quantities of cobalt, at any rate, in plants may be necessary for investigations on cobalt deficiency in animals. There does not seem so far to be any very definite indication that nickel is essential either for any plant or any animal, but since the determination of nickel can be made in the same way as that of cobalt it is as convenient to consider the two elements as cobalt only.

Although these elements can be determined spectrographically, the flame method scarcely has sufficient sensitivity for their ready estimation in plant material, for it would appear from Lundegårdh's data that the smallest amount of either metal measurable in this way is of the order of 100 μg, which means that decidedly large samples of material would generally have to be used.

For the semi-quantitative determination of cobalt in soils Mitchell (1940) recommended the use of the cathode layer arc by means of which 2 p.p.m. of cobalt and 1 p.p.m. of nickel can be detected. The spectral lines used for the determination of these two elements are 3453·5 and 3414·8 Å respectively. For quantitative work an excellent internal standard, particularly for cobalt, is provided by the iron line 3451·9 Å (Scott, 1945, 1946).

Cobalt and nickel can be readily determined simultaneously with the polarograph by the procedure described by Lingane and Kerlinger (1941), the essential feature of which is the use of pyridine as supporting electrolyte. Using a normal solution of potassium chloride containing from 0·05M to M pyridine and 0·05 per cent gelatin the waves of nickel and cobalt are well separated and defined, while manganese does not interfere with them. Ferric iron, if present in large excess, interferes with the determination of nickel and cobalt, but its effect can be eliminated by the use of a supporting electrolyte with pH of about 5·4 containing equal concentrations of pyridine and pyridium chloride in which the ferric iron is precipitated as hydrous ferric oxide. Small amounts of copper do not interfere with the determination of nickel and cobalt, but if present in considerable excess the bulk of it must be removed before polarographing for nickel and cobalt. The author is not aware of the polarographic method having been used for the determination of nickel or cobalt in plant material, and Piper (1942 b) points out that the nearness of the deposition potential of zinc to that of cobalt, and the small amount of the latter relative to that of zinc usually present in plant material, renders the polarographic determinations of cobalt in plant ash uncertain.

The estimation of cobalt in plant material and in soil is usually effected by colorimetric or absorptiometric means depending on the intense coloration produced by cobalt compounds on treatment with the sodium salt of 1-nitroso-2-naphthol-3:6-disulphonic acid, generally known as nitroso-R-salt. Procedures have been described by Kidson, Askew and Dixon (1936) and by Davidson and Mitchell (1940) for the determination of cobalt in soils in this way, by McNaught (1938), Kidson and Askew (1940) and Marston and Dewey (1940) for the similar estimation of this element in plant material, and by Fujimoto and Sherman (1950) for the estimation of cobalt in both soils and plant material. It would appear that quantities of cobalt down to about 0·5 μg are determinable in this way.

3. The Diagnosis of Mineral Deficiencies of Plants

It is obvious that a ready means of diagnosing deficiencies of the various mineral constituents of plants is likely to have great economic value, particularly where crop plants are concerned. Where the deficiency of a particular element is great the plant generally displays symptoms which are readily recognizable by an observer with experience of the effects on the species in question of deficiency in that element. These symptoms are often so definite that the resulting condition has a descriptive name, and a number of well-defined deficiency diseases of crop plants are known to agriculturists and horticulturists; the most important of these are described in a later chapter.* But by the time unmistakable symptoms of deficiency have shown themselves it may be late, and perhaps too late, to effect a cure of the condition by the application of the deficient mineral; this is likely to be so with annuals such as cereals and leguminous crop plants with a short life period rather than with perennials such as fruit and other trees, where the longer life of the plant may provide adequate time for recovery. But even for the latter early diagnosis is obviously desirable in order to avoid a period of feeble growth or poor fruit yield. Also it may be possible that the deficiency of a particular element is insufficient for actual symptoms of a deficiency disease to develop and yet sufficient to bring about a reduction in the rate of growth and finally in crop yield.

A second way of determining micro-nutrient deficiency is provided by analysis of plant material by the methods described earlier in this chapter. This might provide very definite evidence of the adequacy or otherwise of the quantity of the various nutrients in the plant, although it would be necessary first to establish the minimum quantities of the respective mineral elements which must be expected in the different organs of the plants of each species at different stages of development. While

* See particularly Wallace (1943, 1944) for diagnoses and coloured illustrations of crop plants of Britain affected by deficiency diseases, and *Hunger Signs in Crops* by a number of authors (Washington, 1941) for an account, with coloured illustrations, of deficiency diseases of crop plants in the United States.

a certain amount of such information is available it must be admitted that it is far from complete for any one species. The acquisition of the requisite data takes time, but there can be no doubt that the necessary information will ultimately be obtained.

A third method of diagnosing mineral deficiencies, and one which should enable this to be made early, consists in introducing a solution of the salt of the element in question, or even the solid salt, into the plant and observing the reaction. The introduction of the salt into the plant is generally spoken of as 'injection'. The generally accepted meaning of this word is the forcing of material into the organism under pressure, whereas in practice the plant is generally allowed to absorb the solution through a cut surface, or even through an intact leaf, without the application of pressure. However, there is no other simple term to denote this process, and we may follow Roach, who has developed this method for diagnosis of trace-element deficiencies, in extending the use of the term 'injection' to include 'the introduction by various methods of liquids and solutions into plant organs, whether under pressure or not, and their spread therein'.

In a long discussion on injection of plants as a physiological method, Roach (1939) described no less than ten ways in which injection can be carried out; each of these has its own particular value. The ten injection methods are these:

1. Intervenal leaf injection.
2. Leaf-tip injection.
3. Leaf immersion (Anderssen).
4. Leaf-stalk injection.
5. Shoot-tip injection.
6. Branch-tip injection.
7. Shoot injection (Leach) and branch injection (Collison, Harlan and Sweeney).
8. Injection of individual branches.
9. Injection of individual branches together with their roots.
10. Injection of whole trees.

Not all these methods of introducing material into plants have been designed for the purpose of diagnosing mineral deficiencies, nor are all of them equally valuable for this purpose, although any one of them could no doubt serve to demonstrate the existence of such deficiency. But on the whole the methods in which leaves or young shoots are injected are those which are most useful for diagnostic purposes, while those in which larger branches or a whole tree are used are more generally useful for some other purpose, as, for example, the cure of a deficiency.

The principle underlying injection methods of diagnosis is that the introduction into a leaf of a salt of an element in which the plant is deficient will produce a definite response which is in the direction of a cure of the deficiency. The most usual response is a colour change in the leaf which generally becomes greener; sometimes increased rate of growth of a leaf occurs. These responses are best observed when leaf areas permeated by the nutrient are in close juxtaposition to control, non-permeated, areas; a difference in colour between permeated and control areas is then most easily recognizable, while if one simple leaf or leaflet contains both permeated and control areas a difference in the rate of growth of the two parts of the leaf will result in a puckering of the leaf which is readily observed.*

These conditions are fulfilled when leaves of certain species, as, for example, apple, pear, plum, strawberry and broad bean, are subjected to *intervenal injection*. For this treatment a small incision is made near the midrib of the leaf between two major secondary veins. A dilute solution of the salt of the element of which a deficiency is suspected is contained in a small tube, and a wick made of filter paper, or, for small leaves, of darning cotton, passes from the solution through the incision into the leaf. By the use of dyes it is shown that the solute diffuses through the whole area between the two secondary veins and the leaf margin before diffusing into neighbouring intervenal areas. The length of time for which injection is allowed to proceed should be such as to give a long boundary between the injected and neighbouring control area, but not so long that the solute diffuses into neighbouring areas. The best time must be found by preliminary trials, for it varies with the species and climatic conditions. However, for apple, pear, strawberry and Shasta daisy, Roach suggested a period of from 7 to 12 h. The leaves which give the best response are those about half-grown. The maximum response is generally

* It should, perhaps, be pointed out that these responses should only be regarded as indicative of deficiencies when they have been correlated with successful curative treatment, since an improved appearance of the leaf might result from injection without there being a deficiency.

given in about 10 days, but a response has been observed in as short a time as 2 days. This was recorded by Roach as having been observed by Lal as a result of intervenal injection of soya-bean leaves with a 0·025 per cent solution of ferrous sulphate.

In *leaf-tip injection* the tip of a leaf or leaflet is cut off at right angles to the midrib and the cut edge of the leaf immersed in a solution of the substance to be injected. This method can be used for any type of leaf but is particularly suitable for long, narrow leaves. The greater the proportion of the leaf removed, the greater is the penetration of the solute into the rest of the leaf. For example, it was found that if the removed tip contained one-tenth of the midrib, half the rest of the leaf was permeated, but if more than one-fifth of the midrib was contained in the part removed, the whole of the remainder of the leaf and parts of neighbouring ones became permeated. This should be avoided since neighbouring leaves can serve as a control. With compound leaves, such as those of the strawberry, injection can be so contrived that one leaflet becomes permeated and another unaffected, while the third is partially permeated. Roach found that with leaves of apple and pear injection should proceed for about 10 h; a response is apparent in from 7 to 10 days.

With *leaf-stalk injection* the whole, instead of part only, of the lamina is removed, and the leaf stalk left attached to the plant is connected with narrow rubber tubing (such as tyre valve tubing) to a reservoir of the solution. As a result certain leaves of the plant become completely permeated, others partially and yet others not at all; the greater the angular distance of any leaf from the injected stalk the less the permeation. In partially permeated leaves the permeated and non-permeated areas are, at any rate in the case of apple, sharply delimited, and such leaves are considered by Roach to be almost ideal for showing differences in colour and rate of growth of affected and control areas. The method would appear to be applicable to a wide range of species.

For *shoot-tip injection* the tip of a shoot is removed and either a small glass tube of solution is attached to the cut end of the shoot with fine rubber tubing if the shoot is rigid enough to support it, or the cut end of the shoot is bent over into a reservoir of the solution. As a result one or more of the leaves on the shoot become permeated.

The remaining methods of injection listed by Roach are of less interest from the point of view of their value for diagnostic purposes, but reference may be made to the methods used by Anderssen and by Storey and Leach, since these were both devised in connexion with work on mineral deficiency of plants. Anderssen's method consisted in bending over the leaf and immersing it in a weak solution of copper sulphate containing 0·3 p.p.m. of copper. By this treatment chlorotic leaves of plum recovered their normal green colour in 2 weeks, a result which afforded confirmatory evidence that the pathological condition of the trees bearing the leaves was due to a deficiency of copper. Anderssen's work is referred to in more detail in the next chapter.

Storey and Leach (1933) were interested in a disease of the tea plant known as 'yellows', which, as the name implies, involves a chlorosis of the leaves. They traced this to a deficiency of sulphur. Among other pieces of evidence which led them to this conclusion was the effect of introducing various salts into plants growing in the field. The injection of any particular salt was effected by cutting a small side shoot under water and immersing the cut end of the shoot in a solution of the salt. The quantity of solution was maintained by daily additions, and every fourth day the immersed shoots were cut farther back to give a fresh absorbing surface of unchoked wood. It was found that when a 0·5 per cent solution of sodium sulphate, potassium sulphate or magnesium sulphate was used the normal green colour was regained by the leaves on the branch beyond the cut shoot, but that no such recovery resulted when other salts, such as chloride or nitrate, of these metals were used.

CHAPTER III

TRACE-ELEMENT DEFICIENCY
DISEASES OF PLANTS

In chapter I it was noted that many species have been shown to be dependent for growth on one or other of the micro-nutrients, or at least have benefited by treatment with a micro-nutrient. It has been established that certain well-recognized pathological conditions met with in the field are associated with deficiency of a micro-nutrient, and some of these are so widespread or of such economic importance that they are designated by common names. Such, for example, are the grey speck disease of oats, and heart-rot of sugar beet. In this country diseases due to a deficiency of manganese and boron are both widespread and of economic importance; elsewhere shortages of zinc and copper have been shown to be responsible for diseases causing considerable damage to fruit crops. Well-defined diseases attributable to lack of molybdenum have also been recognized.

The more important of the deficiency diseases attributable to shortage of trace elements are described in this chapter.

1. DISEASES ATTRIBUTABLE TO A DEFICIENCY OF MANGANESE

The most general effect of manganese deficiency appears to be in the first place the development of small chlorotic patches localized in intervenal areas of the leaves. The form these patches take in different species is no doubt largely dependent on the anatomy of the leaf, so that in grasses with their parallel venation they tend to take an elongated form, producing 'stripes' or 'streaks', while in reticulate-veined dicotyledons they produce a spotted, speckled or mottled effect, as in potatoes and sugar beet. Other symptoms may follow, including reduction or cessation of growth and the development of necrotic areas which may not be limited to the affected regions of the leaf, but which may even affect the seeds, as in the case of the garden pea.

Grey speck of oats. The disease of oats most usually known as grey speck, but also sometimes called grey stripe, grey spot, or dry spot, is characterized by the appearance in the leaves of spots of a greyish colour, small chlorotic areas, chiefly in the lower half of the leaf, which tend to coalesce and form elongated streaks which finally turn brown. The first sign of the disease often occurs in young plants in the third or fourth leaf. Very characteristically a line of withering and weakness develops transversely across the leaf blade so that the distal portion of the leaf hangs down (see Plate 1 A and B). In the young leaves this line of weakness is often about 1 or 2 in. from the base of the leaf lamina, but correspondingly higher up in older and longer leaves. The leaves may eventually turn completely brown and die. Colour photographs of oats badly affected by grey speck are given by Wallace (1943, p. 95, Pl. 77; 1944, p. 37, Pl. 188).

Badly affected plants may be stunted and die early; in less severe cases flowers may be produced but little grain is formed. Root development tends to be poor, so that affected plants are much more readily pulled out of the soil than healthy ones.

Grey speck appears to be widely distributed. It occurs in different parts of Europe, including Britain, and in America and Australia. It appears to be most liable to occur on certain soils with an alkaline reaction, especially if they contain much humus, and in such conditions the disease may be so serious as to lead to the complete failure of the crop.

It has been recognized for many years that grey speck disease could be controlled by treatment with a soluble manganese salt, either as a soil dressing or by spraying the foliage, but the proof that grey speck was actually related to manganese deficiency was provided by Samuel and Piper (1928, 1929). They grew Algerian oats in carefully controlled water cultures, using carefully purified materials, and with various amounts of manganese sulphate added to the culture solutions containing the usual major nutrients. The initial concentrations of manganese in the different cultures were 0, 1 in 50×10^6, 1 in 10×10^6, 1 in 5×10^6 and 1 in 1×10^6. Cultures grown in solutions free from manganese developed the symptoms of grey speck in about 4 weeks, and this occurred whether the solutions also contained some other trace element such as boron, zinc, cobalt, copper, etc., or not. With culture solutions containing 1 part of manganese in 50×10^6, the symptoms developed suddenly in about 8 weeks. On renewing the culture solution, including the manganese supply, new healthy growth took place, but the symptoms of deficiency again appeared after about 4 weeks. Recovery again took place after a second renewal of the solution. With culture solutions containing 1 part or more of manganese in 10×10^6 no symptoms of grey speck appeared, the solutions being renewed after 10 weeks.

Reference has been made above to the fact that grey speck tends to occur on plants growing on alkaline soils containing much humus. This has led to suggestions that the disease might be associated with excess of calcium ions or with certain organic compounds in the soil. Water-culture experiments carried out by Samuel and Piper to test these possibilities yielded no support for such views. Culture solutions containing calcium ions in various degrees of excess, or various organic substances (humus, sucrose, glucose, starch, cellulose), in no case induced symptoms of manganese deficiency in oats growing in them provided manganese sulphate had been added to the solutions.

That grey speck disease is the direct effect of manganese deficiency was disputed by Gerretsen (1937). He pointed out that Lundegårdh (1932) had recorded that the manganese content of affected plants might be higher than that of healthy ones; indeed, he gave values up to 420 p.p.m. of manganese in affected plants and down to 1 p.p.m. in healthy plants. This is certainly contrary to general experience, and Samuel and Piper found that about 14 p.p.m. was the minimum amount of manganese likely to be present in healthy Algerian oats at the flowering stage.

Gerretsen stated that when a soil which had borne a crop showing symptoms of grey speck was sterilized with formalin, oat plants subsequently grown on it were free from grey speck, although the manganese content was the same as before and there had been no increase in either water-soluble or exchangeable manganese. On re-infecting such sterilized soil with 10 per cent of the original soil grey speck again appeared in oats grown on it and the dry weight of the plants was reduced to 59 per cent of that of plants grown on the sterilized soil, while the manganese content of the affected plants was reduced from 51·5 to

19·3 p.p.m. Some very striking results were given by sand-culture experiments. Plants were grown in sterile sand and in the same sand infected with 5 per cent of so-called 'diseased' soil. The plants in the sterile sand were all healthy, had a mean dry weight of 436 mg and a manganese content of 15·0 p.p.m., whereas those in the infected sand all showed typical symptoms of grey speck, a mean dry weight of only 216 mg, but a manganese content of 26·6 p.p.m.

Oats grown in sterile water-culture solutions containing very small quantities of manganese showed no symptoms of grey speck, although the plants were stunted and might contain less than 10 p.p.m. of manganese. But when the solutions were inoculated with a root tip from an affected plant or with bacteria isolated from affected roots, the symptoms of grey speck developed strongly. Again, the addition of 0·001–0·002 per cent of Germisan, a germicide, to the culture solution of non-sterile plants kept the plants healthy even when the manganese content was low, whereas without the addition of the Germisan the plants were badly affected.

These facts were held to indicate that grey speck is related to the presence of micro-organisms. According to Gerretsen the roots of affected plants always show signs of microbiological disintegration. It was suggested that alkaline products are produced in the roots by the infecting micro-organisms and that these products are carried in the transpiration stream to the leaves, where they produce the grey spots.

Gerretsen therefore concluded that it is necessary to distinguish between the direct physiological effect of manganese deficiency which is a retardation of growth, and the symptoms of grey speck disease which are related to the infection of the roots by micro-organisms. The capacity of the root to resist parasitic attack by micro-organisms was indeed held to depend on the manganese content, but if the roots were maintained sterile healthy plants were produced in presence of a very small supply of manganese so that the manganese in the plant was only from 5 to 35 p.p.m.

It must be admitted that Gerretsen has made a strong case for the view that grey speck disease is not the result of manganese deficiency only. The problem is clearly deserving of further study.

In wheat Gallagher and Walsh (1943) observed that the first sign of manganese deficiency usually appears with the development of the third or fourth leaf. Frequently there is a general similarity with grey speck of oats, but the transverse line of withering tends to occur nearer the tip of the leaf than in oats, later extending to the lower regions of the leaf. The withering may develop as in oats by the coalescence of small grey elongated areas or it may begin at the leaf margin. As soon as the line of withering reaches right across the leaf the upper part of the leaf soon loses its green colour. Sometimes, instead of the development of the line of withering across the leaf as in oats, small grey oblong areas appear scattered parallel with the veins. The whole plant becomes chlorotic and the leaves wither, beginning at the tip.

In barley Gallagher and Walsh recorded the first symptom of manganese deficiency to be a localized paling of the leaf, followed by the development in a few days of small grey oblong spots with brownish margins. These enlarge and coalesce to form stripes parallel with the veins. Sometimes the spotting is most marked near the tip, which finally withers. The plant as a whole becomes somewhat chlorotic.

Rye appears to be affected by manganese deficiency in much the same way as oats, the spots that develop on the leaves being described by Gallagher and Walsh as whitish. The leaves bend over in the same manner as those of oats affected by grey speck. Neither barley nor rye, however, appears to be so badly affected by manganese deficiency as oats.

From the work of Pettinger, Henderson and Wingard (1932), manganese deficiency appears to produce a chlorotic condition in maize very similar to grey stripe. In sand-culture experiments with maize they met with three types of chlorosis which they attributed respectively to deficiency of magnesium (type A), excess of sodium (type B) and deficiency of manganese (type C). In type A the chlorotic areas take the form of long narrow streaks more or less continuous from the base to the apex of the leaves, but with very irregular margins. In type B the streaks are also continuous throughout the leaf but have very regular margins and occupy the whole intervenal region. In the type attributable to manganese deficiency, on the other hand, the chlorotic areas form discontinuous spots or stripes. White or chlorotic spots first appeared in the cultures when these were about 3 weeks old. As the leaves grew the spots also increased in area, and as the affection became more severe the spots tended to coalesce into elongated chlorotic streaks. The tissue in the middle of the chlorotic areas then turned brown, broke down and was dead. Sometimes the dead tissue fell out of the leaf, leaving a number of holes. The similarity of the condition to grey stripe of oats is striking, and a photograph of maize leaves affected by their type C chlorosis published by Pettinger, Henderson and Wingard bears a close resemblance to that of oat leaves affected by grey stripe reproduced in Plate 1 B.

Pahala blight of sugar cane. The disease of sugar cane named Pahala blight, after a small town in Hawaii where it was first observed, is characterized by a partial chlorosis of the leaves, the chlorotic areas taking the form of long white streaks. These are limited to the leaf blades and do not occur on the leaf sheaths. The third, fourth and fifth youngest leaves are generally those most affected. As the chlorotic cells die red spots appear, and as these increase in number neighbouring spots may coalesce so that continuous red streaks result, and there may then follow splitting of the leaf along the line of the streak. By the time red spots appear the plant is generally very much stunted. A fungus, *Mycosphaerella striatiformans*, frequently appears on the red spots, and when the disease was first described in 1906 it was attributed to the attack of this fungus, but in 1928 Lee and McHargue produced evidence that the Pahala blight results from a deficiency of manganese, the fungal attack being secondary to this. This conclusion is based on three lines of evidence derived respectively from the results of the application of solutions, and particularly powders, containing manganese sulphate to the leaves, from chemical analyses of normal and affected leaves, and from sand-culture experiments.

As regards the effect of applying manganese sulphate to the leaves it was found that this salt, generally applied as a dust with dusting sulphur as a carrier, brought about good recovery of affected plants so that new leaves were healthy and of a dark green colour. No recovery resulted from similar treatment with ferrous sulphate, which indeed had the effect of 'burning' the leaves.

Chemical analyses of normal, semi-chlorotic and chlorotic leaves showed no very marked difference in the content of any of the mineral constituents except manganese. The difference in this element, however, between normal and affected leaves was marked, the percentages of manganese in normal, semi-chlorotic and chlorotic leaves being respectively 0·003, 0·0005 and a trace.

The sand-culture experiments were carried out with cuttings of a variety of sugar cane very susceptible to Pahala blight. Manganese-free sand was used. Ten cultures were supplied with a culture solution free from manganese while another ten, supplied with a solution containing manganese, served as controls. These latter plants grew normally, but the plants grown without manganese gradually developed Pahala blight, and by the end of six months the affection was quite severe.

It may be noted that Lee and McHargue reported that Pahala blight only appeared to occur on plants growing in alkaline or neutral soils. Addition of substances such as sulphur and super-phosphates which increase the hydrogen-ion concentration of the soil and so make the manganese in the soil more readily available for absorption tended to diminish the incidence of Pahala blight.

Speckled yellows of sugar beet. The disease of sugar beet known as speckled yellows also involves an intervenal chlorosis in the leaves, the general appearance of the plant resulting from the presence of the yellow chlorotic areas being indicated by its name. As the disease progresses the margins of the affected leaves curl upwards and over the upper surfaces of the leaves. Other cultivated varieties of *Beta maritima*, namely, mangold, red beet, and spinach beet, may also show the same condition, although in red beet the characteristic speckled yellow effect is masked by the red pigment present in the sap of the leaf cells. Spinach (*Spinacia oleracea*), which belongs to the same family as beet (Chenopodiaceae), may be similarly affected. Some good colour photographs of both sugar beet and red beet affected with the disease are given by Wallace (1943, pp. 97–9, Pls. 81–5).

The attribution of speckled yellows to a deficiency of manganese is chiefly based on the fact that the disease is cured by applications of a soluble manganese salt. Analyses carried out by the writer and Dr K. W. Dent, however, show a very striking difference in the manganese content in normal and affected plants of sugar beet. The affected plants examined were grown by Mr W. Morley Davies on soil which had been heavily limed in order to induce manganese deficiency. The normal plants were grown on an adjoining plot not subjected to heavy liming. The differences in manganese content of healthy and affected plants are clearly shown by the data in Table 7. The values marked p were obtained by the polarograph, those marked a by the absorptiometer.

TABLE 7. *Manganese content of normal and speckled sugar beet*

Date of collection of material	Plant organ	Manganese content in p.p.m. dry matter			
		Normal		Speckled	
		p	a	p	a
10 July 1942	Leaf	181	183	13	10
	Petiole	—	7	—	Not recognizable
	Root	—	6	—	Not recognizable
10 August 1942	Leaf	551	—	51	54
	Root	32	36	18	14·5

Although the determinations by the two methods show in some cases a little divergence, they make it clear that the plants affected with speckled yellows contain considerably less manganese than normal plants. This is particularly so in the leaves, where the manganese content of the speckled plants is of the order of one-tenth that of healthy plants. The data afford supporting evidence for the view that speckled yellows is a manganese-deficiency disease.

Marsh spot of peas. The symptom of the disease of peas known as marsh spot is the occurrence on the seeds in the pod of brown or black spots or cavities on the internal surface of the cotyledons. These necrotic spots generally only affect cotyledonary tissue, although occasionally the plumule may be affected. Pods containing exclusively healthy seed, and pods containing only diseased seed may occur on the same plant, and pods are even found containing both healthy and diseased seed. Externally the plant may appear quite normal, although sometimes mild chlorosis or mottling of the younger leaves may be present. In this country it occurs particularly in Romney Marsh where it appears to be limited to alkaline soils (Heintze, 1938). In

Holland, Ovinge (1935) generally found it on alkaline and relatively new polder soils.

The fact that he found peas affected with marsh spot growing near oats which had developed grey speck led Pethybridge (1936) to suspect that marsh spot might be attributable to the same condition as grey speck, that is, to manganese deficiency. This view was supported by the finding of Löhnis (1936) that peas affected by marsh spot contained somewhat less manganese than healthy peas, and by the experiences of Ovinge (1938) in Holland and of Lewis (1939) in this country who found that the application of soluble manganese salt either as a soil dressing or a spray was effective in reducing the incidence of marsh spot. Also Heintze found that among the Romney Marsh soils those on which marsh spot occurred contained less salt-soluble manganese (manganese extracted by a normal solution of magnesium nitrate or calcium nitrate) than those on which peas were free from the disease, the difference in the extractable manganese being related not to the total manganese content, but to the acidity or alkalinity of the soil.

Determinations of the manganese content of different parts of healthy peas and of peas affected by marsh spot made by Glasscock and Wain (1940) show a considerably lower manganese content in the diseased seed. The peas examined were of the variety Harrison's Glory, the diseased sample having been obtained from Romney Marsh and the healthy peas from Folkingham in Lincolnshire. The results are summarized in Table 8.

TABLE 8. *Manganese content of healthy and marsh-spotted peas.* (Data from Glasscock and Wain)

Part of seed	Manganese content (p.p.m.)	
	Healthy seed	Diseased seed
Germ	15	3
Cotyledon outer tissue	11	5
Cotyledon centre tissue	6	< 2
Seed coat	4	2

The relationship of marsh spot to manganese deficiency was definitely established by Piper (1941) by means of carefully controlled water-culture experiments in which specially purified media and carefully regulated amounts of manganese were used. In addition to the ordinary major mineral nutrients the culture solution contained small amounts of boron, copper, zinc and molybdenum as well as sodium chloride. One series of peas was grown in this culture solution without manganese, to four other series 5, 10, 20 and 500 μg manganese per litre were respectively added. For the first 39 days after the seeds were put to germinate no differences were observable in the various cultures, but then in the manganese-free cultures there appeared mottling of the younger leaves and brown lesions on the internodes and tendrils. In a further 2–3 weeks growth stopped.

In the cultures supplied with 5 μg of manganese per litre these same symptoms appeared, but not until 8 weeks from the beginning of germination. On renewal of the solution, including the manganese, healthy new growth was resumed, but in a fortnight the same pathological symptoms again appeared. The cultures produced a few flowers, but no fruits formed.

The cultures supplied with 10 μg of manganese per litre showed no unfavourable symptoms after 8 weeks, apart from slight mottling of the upper leaves, and after renewal of the culture solution growth was vigorous and moderate flowering took place. After another 3–4 weeks, however, the symptoms of manganese deficiency appeared, and although some fruits formed only a few ripened and these were small and imperfectly developed, while the seeds they contained were all badly affected with marsh spot.

The cultures supplied with 20 μg of manganese per litre grew normally, flowered freely and produced numerous fruits with a good yield of ripe seeds. But although the vegetative parts of the plant were free from symptoms of manganese deficiency, 33 per cent of the seeds were severely affected with marsh spot, 24 per cent were slightly affected, while 43 per cent were normal.

The cultures supplied with 500 μg of manganese per litre grew normally and vigorously and showed no symptoms of marsh spot.

The necessity for manganese was also shown by the yield of the cultures; Piper's results are given in Table 9.

These results demonstrate very clearly that marsh spot arises from a partial deficiency of manganese.

TABLE 9. *Yield of peas in water cultures supplied with different quantities of manganese.* (Data from Piper)

Concn. of Mn (μg per litre)	Yield (g) per plant				No. of seeds per plant	Incidence of marsh spot (per cent)
	Shoots	Roots	Seeds	Total		
0	4·3	1·3	0	5·6	0	—
5	12·5	4·0	0	16·5	0	—
10	18·2	3·7	0·4	22·3	6	100
20	21·9	3·2	10·2	35·3	68	57
500	30·7	3·5	13·5	47·7	88	0

A condition similar to marsh spot was recorded by Löhnis in 1950 as occurring in a variety of the dwarf bean grown on certain soils in Holland while Hewitt had earlier (1945) induced the condition in the broad bean (*Vicia faba*) and the runner bean (*Phaseolus multiflorus*) by growing the plants in a sand culture deficient in manganese.

Frenching of tung trees. *Aleurites*, a genus of the Euphorbiaceae, contains five species which are of economic importance on account of the oil yielded by their fruits. They are all trees growing to a height of 25 to 40 ft. The species *A. fordii*, the tung tree or tung-oil tree is the source of tung oil, a drying oil used in the manufacture of paints, varnishes and linoleum and for waterproofing.

In plantations of tung trees in Florida, Reuther and Dickey (1937) reported a rather widely distributed affection which they described as 'frenching'. It is possible that this disease was previously unnoticed because it was masked by another, known as bronzing, attributed to a deficiency of zinc and described later. In trees affected with frenching, chlorotic areas develop between the veins of the leaves, and as the disease advances the tissue in the chlorotic areas dies and necrotic spots arise. Premature abscission of leaves may follow.

Frenching may also occur in another species of *Aleurites*, *A. montana*, the mu-oil tree.

It was found that frenching was not limited to alkaline soils but was related to a low value of exchangeable manganese in the soil.

In the earlier stages of chlorosis recovery could be brought about in from 3 to 6 weeks by dipping the shoots in a 1 per cent solution of manganese sulphate containing 1 per cent calcium hydroxide and 1 per cent calcium caseinate spreader.

Determinations by Reuther and Burrows (1942) of the photosynthetic activity of affected leaves and leaves which had regained their normal colour by treatment with manganese sulphate did not indicate any very significant increase in photosynthetic activity as a result of treatment, and it is suggested that an environmental condition such as high leaf temperature, solarization or stomatal closure might limit the rate of photosynthesis under field conditions in Florida. Reuther and Burrows also point out that trees severely affected by frenching tend to produce small leaves, so that the total photosynthetic activity of the tree and consequently its production of new material may be reduced by frenching.

2. Diseases Attributable to a Deficiency of Zinc

In America plant diseases attributable to a deficiency of zinc may be serious. They largely affect fruit trees, but have also been recorded as occurring in maize and some other herbaceous plants. Apple and pear trees suffering from zinc deficiency were recognized in England by Bould, Nicholas, Potter, Tolhurst and Wallace (1949).

As with manganese deficiency, the first sign of zinc deficiency is usually an intervenal chlorosis, but in trees this is generally followed by very characteristic symptoms of abnormal growth known as rosetting. In spring, instead of the development of elongated shoots with normal-sized leaves distributed along the length of the shoot, there develops a rosette of small stiff leaves. According to the species affected, the disease is variously known as rosette, little leaf, mottle leaf or yellows. Chandler (1937) preferred to class all these conditions, including those of zinc deficiency in maize, as one disease, which he called zinc-deficiency disease, although he pointed out that the evidence might not yet be sufficient to justify the conclusion that the disease is due only to a shortage of zinc. This attitude is no doubt logically sound, but as the symptoms in different species may vary, and as the investigations of the diseases in the various species or groups of species have been to a large extent carried out independently, it has been considered more satisfactory here to describe the disease in these various groups separately.

Knowledge of the internal symptoms of zinc deficiency is due to the work of Reed and his co-workers. In 1935 Reed and Dufrénoy described the result of a microscopical examination of mottled leaves of *Citrus*, which, as we shall see later, may suffer from a deficiency of zinc. In such leaves the palisade cells are broader than in normal leaves, being often transversely divided so that the cells are rhomboidal rather than columnar in shape, while the contents show various abnormalities. Thus chloroplasts are few, their stromata are often rich in fat, and the starch grains within them are generally thin and elongated. The vacuoles of the cell contain phenolic material and little spheres of phytosterol or lecithin. These substances are absent from normal leaves and tend to disappear when zinc is applied to plants affected by mottle leaf.

Later the cytology of the leaves of a number of other species suffering from zinc deficiency was examined by Reed (1938). These included apricot, peach, tomato, maize, squash, mustard and buckwheat. The general effect of zinc deficiency on the growth of the leaves appears to be retarded differentiation, the palisade cells appearing rhomboidal in shape rather than columnar, while the mesophyll is markedly compact, owing to a great reduction in intercellular spaces. Hypertrophy of cells may also occur, and it may be said that zinc deficiency promotes enlargement of the palisade cells rather than their multiplication and differentiation, while in tomato actual atrophy of mesophyll was observed.

In very young apricot and peach leaves the protoplast shows an abnormally great affinity for dyes. This character disappears later, at any rate in apricot, but proteolysis of the cytoplasm may reduce this to an almost invisible layer.

The chloroplasts are particularly affected by zinc deficiency, for this may result in inhibition of their development or in their destruction, injury being greatest in cells receiving the strongest illumination. Plastid injury may be very localized, affected and normal cells being found in juxtaposition.

The phenolic substances, which were noted by Reed and

Dufrénoy as occurring in zinc-deficient *Citrus* leaves, were also observed in zinc-deficient leaves of apricot, peach and buckwheat, but were absent from similarly affected leaves of mustard and maize. Since some phenolic material is present in the normal healthy leaves of some species such as apricot, Reed concluded that the differences in the content of phenolic substances in healthy and affected leaves may be one of degree. No toxic effect appears to be involved.

Reed (1939) also examined the structure of zinc-deficient leaves of tomato grown in water culture. Such leaves exhibited dwarfing, paleness, downward curvature of the leaflets, incurved laminae and necrotic spots on the midrib and laminae. The palisade cells were longer and the spongy tissue more compact than in normal leaves. The chloroplasts of the palisade cells of affected leaves were small and tended to aggregate at the lower end of the cell and, owing to degeneration of some of them, the number of plastids was abnormally low. Degeneration was even more conspicuous in the spongy tissue, the signs of it being increase in the amount of calcium oxalate, shrinkage, the formation of a melanotic substance, and reduction in size and number of plastids.

Reed has also examined the cytological effects of zinc deficiency in the apical buds of apricot and peach trees suffering from little leaf. In apricot some of the meristematic cells in such buds exhibit strong staining with haematoxylin and methyl green; this is followed by premature vacuolization and polarization. A similar state of affairs was observed in zinc-deficient peach buds except that the strong affinity for dyes was not evident. In the apricot, nuclei may become masked by densely stained masses of cytoplasm, and phenolic materials arise from altered cell constituents. These changes were observed while the buds were still in the resting stage. During the early spring tannins, which are present in normal cells, become replaced by phloroglucinol in affected cells, especially in the more active of these. As growth and differentiation proceed tannin compounds reappear, their accumulation being associated with enlargement of the cells and inhibition of cell division. At the same time the amount of phenolic compounds diminishes, reaching a minimum in early summer, after which their quantity increases until it reaches a maximum at the onset of the resting stage. The accumulation of these phenolic compounds in the vacuoles results in an increase of cell size but does not appear to be connected with necrosis of the cells.

Reference has already been made (p. 3) to the observation of Sommer on the necessity of zinc for the completion of the normal life-cycle of beans and buckwheat. Subsequently, by means of carefully controlled water cultures of garden pea, wax bean and milo (*Andropogon sorghum*), Reed (1942) showed that a supply of zinc was necessary for seed production in these plants. Zinc was supplied in a range of concentrations, namely, 0·0, 0·005, 0·02, 0·10 and 0·20 p.p.m. In garden peas no seed was produced when the concentration of zinc was 0·005 p.p.m. or less, but with zinc concentrations of 0·02, 0·10 and 0·20 p.p.m. seeds were produced, the numbers forming increasing with the concentration of zinc supplied. Results with beans and milo were similar except that in these the minimum concentration necessary for seed formation was 0·10 p.p.m.

Pecan rosette. The pecan (*Carya olivaeformis*), a member of the Juglandaceae, is not cultivated in Britain, but its fruit, resembling a small walnut, was becoming familiar to people in this country in the years immediately before 1939. The tree is largely cultivated in the United States where the disease known as pecan rosette is widely spread, and where, according to Finch and Kinnison (1933), it was recognized by growers as long ago as 1900.

The first symptom of the disease is a yellow mottling of the leaves at the tip of a branch, the chlorosis being often evident as the leaves unfold. The leaves at the top of a tree are generally those first affected. As the disease proceeds the affected leaves remain small and are usually crinkled, brittle and misshapen, while the veins tend to stand out prominently. The chlorotic areas of the leaves are abnormally thin and frequently become dark reddish brown in colour and die. Sometimes the intervenal tissue fails to develop at all, with the result that smooth-margined holes are scattered over the leaf. Internode development is poor and finally the branches die back. The development of lateral buds below the dead region results in the rosette appearance from which the disease takes its name. A morphological examination by Finch and Kinnison of the roots of affected trees revealed no indication of an abnormal condition, either externally or internally.

Death of the tree rarely, if ever, results from rosette, but fruit production may be so poor that the cultivation of the trees becomes unprofitable, and in 1932 Alben, Cole and Lewis stated that in some south-eastern states hundreds of acres of pecan orchards had been abandoned on account of rosette, while in south-western states where plantations were more recent, as many as 95 per cent of the trees were rosetting in some places. There can thus be no doubt of the economic importance of pecan rosette.

Researches by Orton and Rand (1914) showed fairly conclusively that the disease is not due to the attack of any microorganism, nor did it appear to be limited to any type of soil.

At first it appeared that the disease might be related to iron deficiency, for Alben, Cole and Lewis (1932a) found that some improvement in rosetted leaves was brought about by dipping them in, or spraying them with, a 0·6–1 per cent solution of ferric chloride or ferric sulphate.

A little later, however (1932b), they found that favourable results by such treatment were obtained only when galvanized iron containers were used for the solutions. This suggested the possibility that the effect at first attributed to iron might be due to zinc salts present as impurities in the iron salts used, and accordingly treatment with solutions of zinc chloride and zinc sulphate was tried. It was found that an immersion of the terminal branches of trees exhibiting rosette in a solution of an iron salt produced no improvement in the condition of the leaves, but that with a solution of a zinc salt young leaves were restored to their normal condition. Similar favourable results were obtained by the use of zinc-lime and zinc-sulphate sprays. Alben, Cole and Lewis concluded from their experiments that zinc is essential for the healthy growth of the pecan tree.

Similar conclusions with regard to the cause of pecan rosette were reached by Finch and Kinnison (1933). Among other aspects of the problem they examined soils on which rosette appeared, but could find no relation between any soil factor and the incidence of rosette. The effects of a number of substances on affected trees were examined. These substances included salts of iron, magnesium, manganese and zinc. Three treatments were employed: (1) placing the dry material in holes bored in the trunk of the tree, (2) spraying leaves with solutions of the substances or dipping the leaves in the solutions, and (3) injecting the solutions into the tree trunks. With zinc salts a great improvement in the condition of the trees was effected by all three methods of treatment. In some cases, but by no means in all, some improvement was observed with the use of iron salts, a result which could be attributed to the presence of zinc as an impurity in the iron salts, especially as no improvement resulted with the use of purer iron salts. No benefit occurred as the result of treatment with either magnesium or manganese salts.

Determinations of the zinc content in different parts of the terminal 6 in. of some shoots taken from the top of pecan trees were made; the results are summarized in Table 10. They show that the shoots of the tree affected with rosette contained very much less zinc than those of healthy trees, while in a rosetted tree treated with zinc chloride by solid injection and in which recovery from rosetting had taken place there was already after 8 weeks a very considerable increase in zinc content.

TABLE 10. *Zinc content of leaflets, petioles and stems of pecan* (Carya olivaeformis). *The quantities are given as p.p.m. of dry matter.* (Data from Finch and Kinnison)

Condition of tree	Leaflets	Petioles	Stems
Healthy	16·7	11·0	7·9
Healthy	10·0	Trace	10·3
Rosetted	3·5	Trace	Trace
Rosetted; then treated with 6 g zinc chloride injected in trunk (55 days after treatment)	3·9	8·6	15·3

Finch and Kinnison also published data of the zinc content of irrigation waters used in different districts of Arizona for supplying pecan plantations. Five out of six of these waters contained no measurable amount of zinc, and in all cases rosette was severe or common in the districts supplied. In the sixth case the water contained a measurable amount of zinc (0·14 p.p.m.) and the district was essentially free from rosette.

The evidence presented by Finch and Kinnison thus supports the view that zinc is an essential element for the growth of the pecan, and that when there is a deficiency of it the condition known as rosette results. The favourable effect of zinc in controlling pecan rosette was also recorded by Demaree, Fowler and Crane (1933) working in Georgia.

Another member of the Juglandaceae, the walnut, may also exhibit the effects of zinc deficiency, but according to Chandler (1937) these do not include rosetting, although the leaves are mottled, crinkled and rather small.

Little leaf or rosette of deciduous fruit trees. Similar to pecan rosette is the disease of deciduous fruit trees known in California as little leaf. The most characteristic symptom is the development in the spring of rosettes of very small leaves which, according to Chandler, Hoagland and Hibbard (1932), who made a special study of the disease, generally possess less than 5 per cent of the area of normal leaves. The affected leaves generally exhibit a chlorotic mottling. Shoots bearing normal leaves may develop later in the season below the little-leaf rosettes, but as the season proceeds the new leaves are progressively smaller and mottled and may be abnormal in shape. Sometimes after one or two years, in other trees after a much longer period, the branches begin to die back. Fruit generally fails to set on badly affected branches and any which does is small and malformed. Stone fruits tend to have brown areas in the flesh (Chandler, 1937). Chandler, Hoagland and Hibbard mention apple, pear, plum, cherry, peach, apricot, almond and grape as all liable to the affection.

Concluding that little leaf was not the result of attack by micro-organisms, Chandler, Hoagland and Hibbard sought for its cause in the soil. As a result of various fertilizer trials they found that affected trees responded to the application of ferrous sulphate, but only when this contained zinc as an impurity. Further trials showed that it was the zinc that was actually responsible for the improvement. Solid injection of zinc sulphate into the trunks of the trees appeared to be the most effective treatment, but favourable results were also obtained by the use of zinc sulphate as a soil dressing or as a spray in winter. No improvement could be detected as a result of treatment with salts of silver, nickel, cobalt, tin, cadmium, mercury, iron, copper, chromium, manganese, aluminium, molybdenum, selenium, zirconium, uranium, strontium, tungsten and titanium (Chandler, Hoagland and Hibbard, 1934, 1935). A number of organic compounds gave equally negative results.

Determinations of the zinc content of stems and leaves of various fruit trees affected by, and free from, little leaf published by Chandler, Hoagland and Hibbard suggest that the zinc content of shoots from trees in orchards free from little leaf is on the whole higher than in those of trees affected by little leaf. Hoagland, Chandler and Hibbard (1936) were able to induce the symptoms of little leaf in young apricot trees grown in water cultures from which care was taken to exclude zinc, as far as practicable.

Chandler, Hoagland and Hibbard were at first very reluctant to attribute little leaf to actual zinc deficiency. Their reasons for this reluctance were the suddenness with which healthy trees might begin to die from little leaf, the recovery of some trees without any obvious improvement in the zinc supply, and the fact that, whereas trees are susceptible to little leaf, annual plants growing on the same soils are apparently free from any symptoms of zinc deficiency.

That simple zinc deficiency alone may not afford a complete explanation of the cause of little leaf is suggested by the work of Ark (1937). This investigator sterilized, by means of steam, soil from orchards showing little leaf and found this treatment very beneficial to maize and tomato. Also sand cultures of maize were treated respectively with little-leaf soil and sterilized soil. In the former the plants soon showed symptoms of zinc deficiency (white bud, see p. 81), while the plants receiving sterilized soil, or in sand without any soil, remained normal. Further, from little-leaf soils he isolated two strains of bacteria which when added to the artificial culture medium in which maize seedlings were growing induced symptoms of white bud which were removed by raising the content of zinc in the medium or by the injection of a zinc salt into the stems. Addition of one of the bacteria to cultures of peach and walnut also resulted in symptoms very similar to little leaf, symptoms which were prevented by the presence of zinc. Altogether Ark's observations are very reminiscent of those of Gerretsen on the relation of micro-organisms to the grey-speck disease of oats.

Chandler (1937) suggested that if little leaf is the result of a simple zinc deficiency this might be brought about by the production in certain soils of a flourishing micro-flora that absorbs zinc in large quantities. The effect of sterilization, by killing this flora, would thus be to leave more zinc available for higher plants rooted in the soil. Alternatively, Chandler suggested that soil organisms might excrete, or yield on dying, substances which combine with zinc to form insoluble zinc compounds and so render it non-available. In these circumstances the effect of the sterilization treatment might be to break up the insoluble zinc compounds and release the zinc in a soluble form.

Mottle leaf (little-leaf type) or frenching of *Citrus*. The disease of *Citrus* trees known as mottle leaf in California has been described in detail by Johnston (1933). The name is derived from the fact that yellow areas arise between the veins of the leaves giving these a mottled appearance. These chlorotic areas enlarge, and as fresh leaves develop these are progressively smaller until in severe cases they may be only an inch long, with chlorophyll developing only at the basal end of the midrib. In extreme cases dieback may follow, ultimately resulting in the death of the tree. The root system is also affected, the smaller rootlets first and the larger ones later, until in severe examples only large roots with a few rootlets remain functional. All species and varieties of *Citrus* can be affected by the disease. The severity of the affection is increased by extremes of high or low temperature.

Since mottle leaf of *Citrus* and little leaf of deciduous trees often occur in the same orchard, and since Chandler, Hoagland and Hibbard (1932, 1933) had found that application of zinc sulphate was a cure for little leaf, Johnston tried the same treatment for mottle leaf of *Citrus* and found that this resulted in restoring the normal green condition to mottled trees. The zinc sulphate was applied as a circle of salt on the soil round the tree, by injection of crystals of the salt in holes bored in the trunk and then sealed, and by spraying the foliage with various sprays containing zinc sulphate. In all cases a favourable result was obtained. Johnston was of opinion that there were probably several kinds of mottle leaf affecting *Citrus* which were due to different causes, and he proposed to designate the type which responded to treatment with zinc as little leaf, thus bringing the terminology into line with that used for the similar condition found in deciduous trees.*

In Florida the chlorotic condition of *Citrus* trees is known as frenching. This is presumably the same disease as the mottle leaf described by Johnston, and affected trees respond favourably to the application of zinc sulphate. Thus Satsuma orange trees showing symptoms of frenching were treated by Mowry and Camp (1934) with a soil dressing of zinc sulphate, and as a result showed signs of complete recovery.

Whether mottle leaf or frenching is a zinc-deficiency disease in the strict sense is not clear. Johnston, indeed, stated that mottle leaf does not appear to result from soil deficiency and suggests that the zinc may act as an antitoxin.

Sickle leaf or narrow dented leaf of *Theobroma cacao*. In Ghana a pathological condition of cacao ascribed to zinc deficiency has been described by Greenwood and Hayfron (1951). The disease also appears to occur in other parts of West Africa including Nigeria and the Ivory Coast, while a disease described as narrow dented leaf and recorded by Ciferri as occurring in the West Indies (Dominican Republic) and South America (Colombia) appears to be identical with sickle leaf. So far the disease does not appear to be of much economic importance.

The first sign of the disease is a bluish green vein-banding in leaves which, although normal in shape, are dull and leathery. These leaves generally persist undamaged except that when old they may exhibit a marginal scorch. Succeeding leaves may display slight corrugation due to the intervenal spaces becoming convex to the upper surface of the leaf. Sooner or later, often after several weeks, small chlorotic primrose-coloured spots appear on the crests of these convex intervenal areas. These spots may appear singly, about 2 mm in diameter, or may form a line along the crest. The amount of the leaf affected may vary from one or two only of the intervenal areas right up to the whole leaf which then presents a mottled appearance (Plate 5 B). These leaves also persist.

---

* Already in the earlier of the papers by Chandler, Hoagland and Hibbard cited above they had included *Citrus* fruits and walnut along with deciduous fruit trees as liable to little leaf. They also showed (1934) the favourable effect of zinc on orange trees affected in this way.

The leaves of the next growth flush usually exhibit deformity which is already observable in the young leaves before they are a centimetre long. Two types of deformity have been observed. In one type the leaves exhibit a lateral curvature giving the leaf the sickle shape from which the disease derives its name; the convex margin of such leaves may be 33 per cent longer than the concave side. In the second type there is no such lateral curvature but at some distance from the lamina end of the petiole the leaf develops in abnormally narrow fashion and the tip forms a long point which is often curled and with slight crenations. The length may be seven times the breadth as compared with three times in a normal leaf. A certain amount of uneven folding of the intervenal spaces occurs in both types. With the cessation of leaf elongation the veins become dark green and as the leaf ages the paler green areas between the veins become chlorotic, and transparent streaks may appear near the midrib. Like the earlier produced leaves these also persist. In the first flushes to be affected the leaves are of normal size but with subsequent flushes, if the symptoms occur in these, the leaves are significantly smaller.

With later flushes in extreme cases observed in the Dominican Republic chlorosis is more pronounced, intervenal elliptical necrotic patches may occur, the leaf tip may become scorched and the leaf ultimately shed. Death of the terminal bud may follow after which there develop thin secondary shoots bearing small yellowish green lanceolate leaves which are soon shed. The shoots then wither and the plant dies.

Water-culture experiments carried out by Greenwood and Hayfron showed that omission of zinc from the culture solution induced the symptoms of sickle leaf; thus suggesting that the disease results from a deficiency of zinc. This conclusion received support from the results of experiments in which a 0·01 per cent solution of zinc sulphate was injected into the young leaves of affected plants; this treatment brought about the cure of the disease. However, shortage of iron appeared to intensify the effects of zinc deficiency and the greatest success in eliminating the symptoms was obtained by painting the young leaves with a solution containing 0·5 per cent of hydrated ferrous sulphate, 0·33 per cent of hydrated zinc sulphate and 0·05 per cent of sulphuric acid.

In Ghana affected plants were found to occur chiefly on soils containing material from the rotting husks of cacao pods. Such soils are alkaline in reaction and contain an unusually high content of potassium and available phosphate, conditions which tend to reduce the availability of zinc in the soil.

Bronzing of tung trees. The condition of tung trees (see p. 68) known as bronzing was first noted in 1930 by Newell, Mowry and Barnette as occurring in trees in Florida growing on soils containing large amounts of phosphate. Later observation has shown that bronzing is not confined to such soils.

The disease usually appears first in the late spring or early summer or even later. The first symptoms are the appearance of a bronze colour in a number of leaves, together with a deformation of the terminal leaves of the shoots. With the development of the dark bronze colour in the leaves necrotic spots develop and parts of the leaves die so that these appear ragged. Ultimately, a twig may lose many or all of its leaves. After the first appearance of the disease in a tree the severity of attack generally increases rapidly. New leaves are successively smaller and become more deformed, the internodes fail to develop normally, resulting in a bunched appearance of the foliage, the twigs remain thin and adventitious buds sprout on the older wood. Although the affection may be at first local, finally the whole of the tree may be involved. Affected trees are unduly subject to injury by the low temperatures of winter, and a bronzed tree may come into activity in the following spring smaller in size. By the third season after the first appearance of bronzing a tree may be reduced greatly in size and, with its almost bare branches, appear half dead.

Mowry and Camp (1934) found that bronzing could be cured by the application of about $\frac{1}{4}$–$\frac{1}{2}$ lb of zinc sulphate per tree to the soil or by spraying the foliage with a spray containing 6 lb of hydrated lime and 3 lb of 89 per cent zinc sulphate in 50 gal of water containing some calcium caseinate spreader. While not definitely committing themselves to the view that bronzing of tung trees is a deficiency disease they concluded that it should be considered the result of zinc deficiency until proved otherwise. As a result of this work by Mowry and Camp on the control of bronzing by the application of zinc sulphate Reuther and Dickey were able to report in 1937 that comparatively little bronzing was then to be found in properly conducted tung plantations.

Pinus radiata rosette. Trees of *Pinus radiata* growing in plantations on poor soil in Western Australia are liable to exhibit a condition of rosetting reminiscent of that occurring on so many species of trees in America already described. Field trials of the effect of spraying with zinc chloride or zinc sulphate described by Kessell and Stoate (1936, 1938) showed that, as with other species affected by rosetting, treatment with zinc was a cure for the disease. The presumption that rosette of *P. radiata* is brought about by a deficiency of zinc was shown to be correct by carefully controlled water cultures carried out by Smith and Bayliss (1942). Removal of zinc, copper and lead from the stock solutions was effected by the use of dithiocarbazone. Boron, manganese and copper were added to the culture solutions containing the usual major nutrients. Zinc was added to the controls only. The method of purifying the solutions did not remove molybdenum and it was not necessary to add it.

The first symptom of zinc deficiency was a decreased rate of growth which began to be noticeable after about 3 months. This was soon followed by the shoot apices acquiring a flat-topped appearance due to the lower activity of the apical meristem. The apical buds appeared bunched together while secondary needles stopped growing and so appeared short, thick and stiff and the bundles did not spread open. Chlorotic symptoms did not appear until the lapse of another 5 weeks. Then small yellow dots appeared near the tip of the needles followed by browning, soon giving the tops of the branches a bronzed appearance. With the appearance of the first signs of chlorosis in the leaves conical swellings sometimes arose on the root apices. As far at least as the shoots are concerned the effect of zinc deficiency on *P. radiata* thus resembles the effects produced on other trees.

It should be noted that the cultures grown by Smith and Bayliss were non-mycorrhizal.

White bud of maize. As stated early in this book (p. 3), as long ago as 1914 Mazé demonstrated the essentiality of zinc for the growth of *Zea mais*. Mazé reported that maize plants grown in water culture deficient in zinc at first developed normally, but that quite suddenly the leaves darkened and developed a metallic

6   STE

sheen while nocturnal exudation became very abundant, a deposit of soluble salts being left on the leaves. Death of the plants followed in 3–5 days.

According to the investigations of Barnette and Warner (1935) it would appear that the disease of maize known in Florida as white bud is due to zinc deficiency. They reported the disease as occurring chiefly on land which had been under cultivation for a number of years and also on poorer land recently brought into cultivation. The disease may sometimes be serious enough for the crop to fail completely.

The first symptoms of the disease may appear within a week of the emergence of the seedlings from the soil. The first sign of the trouble is the appearance of light yellow streaks between the veins of the older leaves followed by the rapid development of white necrotic spots. As the newer leaves unfold they are often pale yellow to white in colour, a symptom which gives its name to the disease. The older leaves develop light slate to dark brown necrotic areas which increase in size and merge with one another until the whole leaf is dead. Meanwhile the younger chlorotic leaves continue to unroll and as they expand develop typical yellow intervenal striping. Internodes fail to develop properly so that the whole plant is stunted. The root system, however, appears to be normal. Zinc-deficient plants of another cereal, oat, and of broad bean, are shown in Plates 3 and 4.

The effect of various fertilizer treatments on maize plants affected with white bud was examined by Barnette and Warner. They found that a dressing which included 20 lb of zinc sulphate per acre induced a very much higher yield of grain than any inorganic dressing which did not include zinc. With the application of zinc sulphate the chlorotic plants recovered a normal green colour, made healthy growth and produced grain. The same effect was produced by the application of stable manure or leaf mould, while chicken manure and alkaline peat produced some improvement in the condition of affected plants. Spectroscopic examination of the ash of these various organic manures revealed the presence of zinc in all of them. The work of Barnette and Warner, although not definitely establishing white bud of maize as a disease due to deficiency of zinc, indicated the probability of this being the case.

The possible relationship of bacteria to white bud is considered in the discussion of little leaf of deciduous fruit trees (p. 76).

Zinc deficiency in dwarf beans (*Phaseolus vulgaris*). Plants of *Phaseolus vulgaris* exhibiting pathological symptoms attributed to deficiency of zinc have been found growing on a number of soils in the United States. Such soils are some peat soils in Florida, soils in old corral areas in California and freshly irrigated soils in Washington. The symptoms as described by Viets, Boawn and Crawford (1954) consist of first a mild general chlorosis which becomes intensified in the intervenal areas followed by the production of brown spots and necrosis of the mesophyll. Abscission of older leaves and flowers also occurs. Few seeds are set and those which do are small and mature slowly. The view that these symptoms result from zinc deficiency was supported by the results of experiments in the field by Viets and his co-workers with the variety Red Mexican. Plants were supplied with zinc at four levels described as very low, low, medium and high. Symptoms described above were most severe in plants supplied with zinc at the very low level, were still severe in those supplied at the low level, but were absent from those receiving medium or high levels of zinc supply. The effect of the different zinc treatments on the dry weight of the plants was very striking. The dry weights per plant with the four treatments were, with increasing zinc supply, 2·7, 5·7, 13·6 and 15·7 g, while the corresponding amounts of zinc per plant were 0·025, 0·061, 0·251 and 0·544 mg. Zinc-deficiency symptoms were apparent in plants in which the zinc concentration in the mature leaves was about 20 p.p.m. or lower.

3. Diseases Attributable to a Deficiency of Boron

The effects of boron deficiency in plants have been well dealt with by Brenchley and Warington (1927) and more recently by R. W. G. and A. C. Dennis (1937, 1939, 1941, 1943). From the accounts of these and other authors it appears that the first external symptom of boron deficiency is generally the death of the apical growing point of the main stem. This is followed by growth of lateral buds into side shoots, the apices of which then die. Further symptoms are slight thickening of the leaves, a tendency for these to curl, and sometimes a slight chlorosis. The petioles, and even the leaves, often become brittle, flowers may not form, or if they do fruit may not set. Stunted root growth is general.

Quite a number of investigations have been made on the histology of boron-deficient plants. These include broad bean (Warington, 1926), tomato (Johnston and Dore, 1929; Fisher, 1935; Van Schreven, 1935; Palser and McIlrath, 1956), *Citrus* (Haas and Klotz, 1931), tobacco (Van Schreven, 1934), sugar beet (Jamalainen, 1935), maize (Eltinge, 1936), potato (Van Schreven, 1939), apple (MacArthur, 1940), carrot (Warington, 1940), *Brassica* spp. (Chandler, 1941), radish (Skok, 1941), squash (Alexander, 1942), sunflower (Lowenhaupt, 1942), garden beet and cabbage (Jolivette and Walker, 1943), turnip and cotton (Palser and McIlrath, 1956).

In *Vicia faba* Warington found that the chief internal symptoms of boron deficiency were frequent hypertrophy of the cells of the cambium and subsequent discoloration and degeneration of the cells; disintegration of the cells occurred, however,whether there was previous enlargement or not. Disintegration of phloem and ground tissue was also frequent, while xylem development was poor and the cells might also ultimately disintegrate. Brenchley and Thornton (1925) had previously found that development of the xylem of the root nodules either failed altogether or was very poor.

In sugar beet and *Citrus* hypertrophy of the cambium, followed by its disintegration and that of the phloem, is also symptomatic of boron deficiency. In *Citrus* boron was found to be essential for meristematic and cambial activity. In tobacco Van Schreven found that the first symptoms appeared in the cells of the root apex and later in the stem apex. In these regions cells became brown and degenerated. The symptoms then extended backwards from the apices, the vascular tissue being particularly affected. Proliferation of the cells of the phloem occurred resulting in the individual elements being compressed and distorted, while the xylem development was poor. Cells of the ground tissue might also undergo disintegration. The same worker recorded similar effects in tomato where thin-walled cells, such as those of cambium, phloem and ground tissue, undergo degeneration. The degeneration of the phloem in tomato was earlier recorded by Johnston and Dore and at about the same time by Fisher.

A number of species of *Brassica* were examined by Chandler. In broccoli and Brussels sprouts boron deficiency was found to bring about cessation of cell division in the root apex and degeneration of the root cap. In rutabaga the thin-walled cells of the meristematic tissue of the stem apex and of the root cortex became crushed, while cells near the cambium became elongated and the cork cambium failed to produce cork. In cabbage Jolivette and Walker recorded considerable proliferation of cells in the cambial region resulting in a zone of meristematic tissue between xylem and phloem several times the usual width of this band, while there was a corresponding reduction in the amount of differentiated xylem and phloem. In radish Skok found that vascular tissue near the axis which might have been produced while boron from the seed was still available, or from boron present as impurity in the substrate or chemicals used, was usual, but that there was a complete absence of normally developed and lignified xylem vessels in the region between this inner tissue and the cambium, while the later formed phloem disintegrated. Xylem parenchyma cells, although smaller than normal, appeared uninjured. Owing to the failure of the development of normal vascular tissue the roots cracked and then the cambium and phloem mostly disintegrated. In swede, according to Jamalainen, the first symptom of boron deficiency is enlargement of xylem parenchyma.

In the squash Alexander noticed that boron deficiency brought about hypertrophy and collapse of the meristematic cells and of more mature cortical cells of the apical region of the stem, while towards the apex of the root, cells of the central cylinder were similarly affected. In older regions of the stem parenchyma of the ground tissue, xylem, and the region between xylem and internal phloem showed abnormal enlargement. Cells of the cambium were also enlarged. Enlargement of thin-walled cells also occurred in the leaves.

In garden beet the first internal symptom, according to Jolivette and Walker, appears in the phloem where certain cells resembling companion cells become filled with a densely staining substance. Occasional hypertrophy of cambial cells was also noted. Later degeneration of xylem tissue occurs.

The first internal deficiency symptom of maize was found by Eltinge to be a disintegration of some of the leaf cells. Later, entire cross-sections of a leaf might collapse, and in other parts of such leaves cells might fail to differentiate, and in yet other parts hypertrophy of cells, particularly those of the lower epidermis, might occur. Later, disintegration of cells of the stem apex took place.

Thus, although the internal symptoms vary somewhat from species to species, it may be said that in general boron deficiency leads to degeneration of the meristematic tissues, including the cambium, to breakdown of the walls of parenchyma cells and to feeble development of the vascular tissues. Of these the phloem appears to be most affected, but imperfect development of xylem is also a common feature. Hypertrophy of thin-walled cells and then discoloration are frequent precursors of their disintegration. Sometimes this latter is preceded by abnormally active cell division.

Notable differences in the effect on the anatomy of plants as a result of boron deficiency were observed by Palser and McIlrath (1956) between turnip and tomato on the one hand and cotton on the other. In all three species boron deficiency was characterized by smaller vessels in the secondary wood and changes in the collenchyma and sclerenchyma, but the cells which became hypertrophied in cotton were those of the outer region of the pith and in the neighbourhood of the protoxylem whereas in turnip and tomato it was those of the cambial region which were first affected in this way. In turnip these cells in the root, and in tomato those of the upper region of the shoot, and in both species those of the leaf lamina, were the first to become hypertrophied. The hypertrophy of some of the cells was followed by necrosis and cell division occurred in neighbouring cells. The symptoms later spread from the cambial region to the phloem and sometimes to the xylem parenchyma.

Heart rot of sugar beet and mangold. The disease of sugar beet and mangold known as heart rot, crown rot or dry rot, is widely distributed through Britain, Europe and America.

It is generally most severe on alkaline soils and in dry years. As the names given to it imply, the most prominent symptom is a necrosis of tissues of the crown and interior of the root. The first symptoms of the disease, however, appear in the youngest (inner) leaves (cf. Plate 5A). These are stunted, become markedly curled, and the petioles develop a brown to black colour which may extend into the veins of the lower part of the laminae. As the plants grow older the leaves become affected, the veins becoming yellowish and the petioles brittle. Next, the inner leaves become brown or black and die, the main growing point dies and the outer leaves turn yellow, wilt, wither and finally also die. New shoots now develop in the axils of the dead leaves, but these become affected in the same way as the earlier formed leaves.

With the death of the first crop of leaves the tissue of the crown begins to rot. First necrosis occurs at a number of spots on the crown and these develop inwards into the root, increasing in size until in severe cases the greater part of the tissues of the root may be destroyed. Invasion of the affected parts by *Phoma betae* usually follows. The percentage of sugar in even the healthy parts of affected beets is less than that of healthy roots, so that from an economic point of view heart rot can be of very serious consequence.

The connexion of heart rot with boron deficiency was shown by Brandenburg by means of controlled water cultures, first of mangolds (1931) and later of sugar beet (1932). Similar results were obtained by Bobko and Belvoussov (1933) and by Rowe (1936). In the experiments of Bobko and Belvoussov seedlings provided with no boron developed symptoms of heart rot after about a week. Renewed root development of boron-starved cultures resulted on the addition of boric acid to the culture solutions. Too high a concentration proved toxic, the most satisfactory range of boron concentrations being from 0·5 to 5 mg of boric acid per litre. These findings were confirmed with sand cultures and field trials carried out by Brandenburg and others.

Canker and internal black spot of red or garden beet. Pathological conditions of garden beet attributed to boron deficiency have been described both here and in the United States. A full description of the disease as it occurs in the

United States has been given by Walker (1939), who terms it internal black spot in reference to its most prominent symptom, the presence of internal hard black necrotic masses which render the beet unfit for canning. These masses, irregular in size and shape, are not characteristic of any particular part of the root, for although they may be confined either to the central region or the peripheral region, they may also be scattered throughout the root. When the black spot occurs near the periphery a rift in the surface tissues may occur, soil micro-organisms may enter and attack the root and a surface canker may result. But when the necrotic areas are well inside the root there may be no external symptom of the disease.

In this country the lesions on boron-deficient garden beet appear always to be superficial and are known as canker. Thus Wallace (1943) states that 'rotting occurs on the sides of the roots and may not penetrate into the more central tissues', and he gives some good colour photographs of cankered garden beet (*op. cit.* Pls. 106, 107, pp. 109, 110). The absence of the internal necrosis in red beet grown in this country is not explained, but it may be related in some way to varietal differences in growth. Walker states that the necrotic areas are most obvious between the prominent rings marked by the thick-walled vessels formed by the activity of the secondary cambium zones in the pericycle.

The symptoms of boron deficiency exhibited by the shoot of garden beet are similar to those of sugar beet and mangold but generally less conspicuous. The older leaves are generally normal, but the younger leaves towards the middle of the crown become malformed, abnormally small, and rich in anthocyanin. The malformed leaves tend to die early and a rosette of dead, stunted leaves may result. This condition is generally followed by the sprouting of dormant buds at the base of the dead leaves and the consequent development of new leaves, which, however, generally develop the characteristic symptoms indicative of boron deficiency.

Brown heart or raan of swede and turnip. The disease of swedes generally known as brown heart in England and raan in Scotland has a wide distribution, having been recorded in all parts of Britain, different countries of northern Europe, Iceland, Canada, Newfoundland and the United States, Australia and

New Zealand. A good account of it has been published by Dennis and O'Brien (1937).

No external symptoms of brown heart appear during the growing season, and the root appears normal externally. On cutting an affected root across, however, the middle region presents a mottled appearance owing to a discoloration of patches of tissue within the xylem (see Plate 6 A and B). According to Dennis and O'Brien the affected tissue is limited to that within the cambium. Usually it is the outer region of the xylem which is affected so that the middle of the root has a normal healthy appearance, but in severe cases the whole of the region within the cambium may exhibit mottling, and in very severe cases the central tissue may break down and the root become hollow (Wallace, 1943, and see Plate 7 A). The colour of the affected patches shows some variation. Sometimes they have a greyish, slightly brownish or 'water-soaked' appearance (Dennis and O'Brien, 1937; Wallace, 1943), the coloration being partly due to reduction in the intercellular spaces owing to slight swelling of the affected cells, partly to the development of a brown pigment apparently connected with slight swelling of the cell walls. Dennis and O'Brien state that bacteria can always be isolated from the affected tissue, and they suggest that the coloration may be partly related to increased activity of bacteria normally present in the intercellular spaces.

Swedes affected with brown heart are unfit for human consumption, as the discoloured parts remain hard when the root is cooked, while affected roots contain less sugar than healthy roots and have a bitter taste.

In 1934 Güssow reported that application of boron as sodium tetraborate resulted in a large measure of control over brown heart, whereas other elements were not effective, and in the following year a number of investigators working independently in different countries published the results of field experiments on the control of brown heart by the application of borax or boric acid. These workers included O'Brien and Dennis in Scotland, Whitehead in Wales, Jamalainen in Finland and Hurst and Macleod in Canada.

The proof of the essentiality of boron for swedes was, however, provided by the sand-culture experiments of Hill and Grant (1935) and Jamalainen (1935 a, b) and the water-culture experiments of Dennis and O'Brien (1937), which clearly indicate that without a supply of boron swedes fail to grow beyond a very young stage.

The effect of boron deficiency in turnips is, as might be expected, similar to that in swedes.

Browning of cauliflower. The most noticeable symptom of boron deficiency in cauliflower is the formation of brown patches in the heads. The disease has been investigated by Dearborn *et al.* (1936, 1937, 1942) by means of field trials, pot cultures and histological examination. The first external sign of the disease is the appearance of 'water-soaked' patches on the developing head (see Plate 7 B). These patches soon turn brown and hard and in wet weather may rot. A certain amount of chlorosis may become apparent in the leaves, particularly at the apices of the older ones, which may become thicker, brittle and liable to curl downwards. Blistering may occur on the petiole and along the midrib. The root system is poorly developed. The internal symptoms resemble those found in other cases of boron deficiency (see p. 84). Thin-walled parenchyma cells of pith and cortex appear to be the first affected, individual cells undergoing enlargement while the intercellular spaces appear to become filled with mucilage, giving the tissue first a water-soaked appearance, followed by a gradual darkening to a deep brown colour. The brown patches on the head appear to arise in the same way. Very characteristic is the breakdown of the pith to form an elongated cavity (Plate 8).

In field experiments it was found that browning of cauliflower was completely eliminated by a dressing of 10 lb of borax per acre.

Cracked stem of celery. A disease of celery known as cracked stem is widely distributed in the United States and Canada. It was first recorded in Florida and described by Purvis and Ruprecht (1935, 1937). The first external symptoms of the disease are a brownish mottling of the leaf, first in the marginal region, and brittleness of the stem. Next, cracks develop transversely across the leaf stalks, which then curl outwards and turn brown. The roots suffer the same discoloration and finally die. In extreme cases the terminal bud may suffer the same fate. Where the disease is prevalent it may be very serious and bring about a loss of half the crop in bad cases. Investigations by Purvis and Ruprecht, including the use of water cultures and sand cultures, as well as field trials, indicate that cracked stem results from boron deficiency and can be controlled in the field by the application of borax either as a soil dressing or as a spray.

As far as I am aware, cracked stem of celery has not been recorded in Britain, although I have seen examples of it induced artificially by Mr Morley Davies through heavy liming of the soil.

Lucerne yellows or yellow top. The effects of boron deficiency on lucerne (alfalfa) were described by McLarty, Wilcox and Woodbridge (1937) as a uniform yellowing of the terminal leaves, or a bronzing of the intervenal areas, poor development of internodes and death of the growing points. It appears that the symptoms may be confused with those resulting from attacks by the potato leafhopper (*Empoasca fabae*), and Colwell and Lincoln (1942) have published a comparative account of the symptoms produced in lucerne by boron deficiency and by the leafhopper, their account being the result of work carried out both in the greenhouse and under field conditions. From this it would appear that when due to boron deficiency yellowing or reddening is always confined to the terminal leaves, including those of lateral branches, whereas when due to the leafhopper, yellowing or reddening may occur at various levels on the shoots. In boron-deficient plants the terminal internode is always abnormally short, whereas with leafhopper injury this is not so, although the plant may be generally stunted. With boron deficiency the terminal bud is always abnormal, and death of the growing points results. Colour photographs of boron-deficient lucerne are published in the paper by Colwell and Lincoln referred to above.

Several workers have reported increased growth and flowering, and increased yield of seed of lucerne as a result of application of boron to the soil (e.g. Grizzard and Mathews, 1942), although apparently the non-treated plants did not suffer from yellowing. This suggests that boron deficiency may be met with which is insufficient to induce yellowing but which is yet sufficient to bring about a decrease in growth, flowering and fruiting.

Top sickness of tobacco. The effect of boron deficiency on tobacco growing in the field in Sumatra was described by Kuyper (1930), who called the disease 'top sickness'. In America the effect of boron deficiency on tobacco growing in water cultures, pot cultures and on field plots has been described by McMurtrey (1929, 1933, 1935). The first external symptoms appear in the terminal bud, the leaves of which have a pale green colour, the bases of the leaves being paler than the tips. At the same time the leaves have stopped growing. The tissue at the base of the young leaves of the bud now breaks down and the bud dies. Next the older leaves become thicker and increase in area and later become brittle. The midrib may break and the vascular tissue then becomes discoloured. The upper leaves tend to take a drooping position. If the disease is not too severe and lateral buds develop in the axils of the leaves, they generally degenerate in the same way as the terminal bud. If there is partial recovery from the disease the younger leaves and the upper part of the stem as they develop may appear twisted to one side owing to growth round the damaged tissue.

Internal cork of apples. Disease of apples as a result of boron deficiency occurs in Europe, Canada, the United States, Australia and New Zealand. According to Dennis (1937) the disease was recorded in Australia as long ago as 1892, when it was, however, confused with bitter-pit. The symptoms of the disease are rather varied, and partly as a result of this a number of names have been given to it including internal cork, drought or drouth spot, corky pit, corky core, brown heart, poverty pit, die-back and rosette. Internal cork refers to lesions which first appear as clear or slightly greenish rounded regions, generally not exceeding 1 cm in diameter, and which may arise anywhere in the flesh of the fruit. The patches become dry and darken, and finally are dark brown in colour and of a corky or spongy consistency according as the lesions appear earlier or later in the developing fruit (Carne and Martin, 1937; Burrell, 1937). In apples affected with drought or drouth spot the lesions are in the form of large superficial patches, while in the case of corky core it is the middle region including the core which is affected. According to Burrell a greater deficiency of boron is required to produce rosetting and die-back than to induce internal cork in the fruit. Many investigators have demonstrated that this disease can be controlled by the application of borax or boric acid to the soil (see, for example, Askew, Chittenden and Thomson, 1936). Internal cork does not appear so far to have been recorded in Britain.

There is evidence that boron deficiency may also induce the formation of internal cork in the fruit and die-back of the shoots in pears.

Brown-spotting of apricots. On certain light-textured soils in New Zealand apricots tend to develop brown spots in the flesh, especially near the stem end, and a dry, spongy condition round the stone. This condition is attributed by Askew and Williams (1939) to boron deficiency. The condition is controlled by the application of ½ lb of hydrated borax per tree to the soil, or as a 0·1 per cent spray. The increase in the boron content of the leaves and fruit resulting from these treatments was accompanied by freedom from brown-spotting.

Die-back of raspberries. A disease of raspberries first recognized by Askew, Chittenden and Monk in 1951 as resulting from shortage of boron is characterized by the abnormal development of the leaves, retardation of the opening of the buds or even complete failure of the buds to open, and failure of the fruit to set. In the field the abnormal development of the leaves is not usually apparent in the young canes of the current season's growth but in older canes the leaves are long, thin and deeply indented, presenting a feathery appearance. In less severely affected plants leaves curl downwards, their surface is crinkled and the serrations of the margins ill-defined. Most of these symptoms were produced by Monk (1955) in sand cultures without a boron supply. In these the feathery leaf symptom occurred in the young canes and the other symptoms, although these did appear, were not so marked as in plants growing in the field.

Ivy leaf of hops. In commercial hop gardens plants have been observed exhibiting symptoms of what has been described as 'ivy leaf'. The incidence of this appears to be greater in the hop gardens of the West Midlands of England than in those of Kent where it has not been regarded as of economic importance. Evidence that this condition resulted from boron deficiency was provided by Cripps (1956) who examined the effects of boron deficiency in hop plants grown from cuttings in sand cultures deficient in boron. When this was accompanied by a high or medium level of calcium supply the rootstocks first produced many shoots with short internodes about 1 cm long. The growing points of these shoots soon turned brown and died; further shoots which then developed died similarly while many lateral buds on the main shoots died without expanding. What lateral buds on the main shoots did expand developed with very short internodes and these stunted branches finally died back. The whole plant thus had a bushy appearance. The first leaves to expand were very small with slightly chlorotic margins while leaves produced later were small, often not lobed, distorted and deeply toothed. The tendency to produce deeply indented leaves was most marked in boron-deficient cultures supplied with calcium at a low level and the symptom in these plants was very similar to the ivy-leaf condition of plants in the field. Root development of plants receiving no boron was poor in comparison with those supplied with boron to the extent of from 0·5 to 2·0 p.p.m. in the medium. A supply of 10 p.p.m. brought about the development of symptoms of boron toxicity.

4. DISEASES ATTRIBUTABLE TO A DEFICIENCY OF COPPER

Although copper has now been shown to be essential for the growth of a number of plants, diseases attributable to a shortage of this element are rarely met with in the field and are not of significance in Britain. The two well-recognized diseases which appear to be related to copper shortage are an affection of fruit trees known as exanthema, die-back or chlorosis, and a disease of various herbaceous plants known as reclamation disease. These are described later.

The general effects of copper deficiency on the leaves of tomato plants have been described by Reed (1939). The leaves exhibit restricted growth, but although they are considerably smaller than normal leaves the number of leaflets and lobes of the leaflets is not reduced. The laminae of the leaflets are in-curved and they develop a bluish green colour with a distinct sheen. Later necrotic areas appear.

Microscopic examination shows that in the early stages of development the palisade cells of affected leaves contain many large hyperchromatic plastids. These plastids ultimately de-generate, at the same time tending to form aggregations at one end of the cell. Cavities tend to form below stomata by the separa-tion of the upper ends of adjacent palisade cells. The cavity may lengthen so that ultimately it extends the whole length of the cells and at the same time broadens owing to the shrinkage of the cells, this being followed by a disappearance of the cells owing to lysis of their contents. These processes appear to lead to the production of the necrotic areas already mentioned.

Exanthema or die-back of fruit trees. A pathological condition known as exanthema affects various fruit trees in-cluding both *Citrus* species and rosaceous trees as well as the olive. The disease was recorded in *Citrus* in Florida in 1875, but it was not described for other trees until 1928, when Smith and Thomas recorded that for a long time it had been known in California as affecting other fruit trees. They reported it as occurring in French prune, Japanese plum, apple, pear and olive trees.

Exanthema in *Citrus* is of widespread occurrence throughout the world. The symptoms of it in *Citrus* trees in Western Australia were described by Pittman (1936) as follows. In the orange strong water shoots tend to bear abnormally large leaves while the shoots themselves, instead of growing straight, form an S-shaped curve. Small blister-like swellings containing gum develop on the young shoots. Later these swellings develop into longitudinal ruptures bordered with brown or reddish brown ridges from within which the yellow or reddish gum exudes in wet weather. Shoots so affected lose their leaves and die back, and lateral shoots developing at the base of affected twigs produce a typically bunchy appearance. A condition of 'multiple-bud' development, resulting from the development of a cluster of buds instead of two, is frequent, thus emphasizing the bunchy habit of the tree. Die-back is typical of badly affected trees. The fruit is small and frequently marked with irregular-shaped brown spots or blotches where finally the skin, after becoming dry, splits open.

In the lemon, gum formation on the branches is rare, but gum pockets may develop on the skin of the fruit.

Haas and Quayle (1935) stated that leaves of *Citrus* in the early stages of exanthema may be unusually green, but may become mottled or chlorotic in later stages of the disease.

According to Smith and Thomas (1928), in French prune trees (*Prunus domestica*) affected with exanthema there is vigorous growth of new shoots each spring, but in June the terminal buds wither and the terminal leaves develop a chlorosis. There is a similar development of lateral shoots and of multiple buds as in *Citrus* as well as the production of eruptions of the bark. Apples, pears and Japanese plums can develop similar symptoms. Pittman describes exanthema of Japanese plums in Western Australia as involving the development of cracks in the bark which in some varieties may reach the cambium. Later, exudation of gum takes place through the ruptures. Die-back is here also very characteristic of the condition.

As long ago as 1917 Floyd reported the beneficial effect of copper sulphate on *Citrus* trees in Florida affected with this disease, and confirmation of this was forthcoming from Wickens (1925) in the treatment of affected orange trees in Western Australia, from Smith and Thomas (1928) in regard to plum, apple, pear and olive in California and from Pittman (1936) for *Citrus*, plum and apple in Western Australia.

Oserkowsky and Thomas (1933) showed that the condition of the leaves of Bartlett pear trees in relation to this disease corresponded to their copper content. Thus, leaves affected with exanthema contained 3·1–5·1 p.p.m. of the dry weight, while normal leaves from localities free from the disease contained 11–20 p.p.m. of the dry weight. While these workers considered that these analyses afforded strong evidence that exanthema resulted from a deficiency of copper, they pointed out that there was no evidence to decide whether the disease was due directly to copper deficiency or whether the effect was an indirect one such as might be brought about, for example, if the action of copper resulted from its neutralizing the effect of toxins absorbed by the plant from the soil.

More definite evidence that exanthema in *Citrus* is due to a deficiency of copper was provided by a culture experiment described by Haas and Quayle (1935). Valencia orange trees grafted on sour-orange stocks were grown in twelve large tanks containing pure sand. A nutrient solution containing all the known mineral nutrients except copper was supplied to the trees. Although for some years the trees grew vigorously, after seven years they displayed typical and pronounced symptoms of exanthema including the S-shaped growing shoots, exudation of gum, blisters on the surface of the shoot and dying back of many shoots. The leaves became covered on the ventral side by a resinous stain and many leaves developed chlorosis.

The effect of copper deficiency on deciduous fruit trees in South Africa has been described by Anderssen (1932). Apple, pear, plum, peach and apricot trees are all affected to different degrees. Thus plums, peaches and apricots exhibit a very decided chlorosis in the areas of the leaves between the veins. This is accompanied by very marked rosetting of the leaves, cessation of apical growth followed by the dying back of the branches from their apices. Apples, on the other hand, exhibit chlorosis much less frequently, but rosetting is severe and the long shoots die back. Pears similarly do not show chlorosis very often, but unlike apples they do not develop rosetting; however, the shoot apices and the youngest leaves become browned and finally die back.

Anderssen makes no reference to the surface eruptions on the shoot which are so characteristic of copper-deficient *Citrus* trees as to give the name exanthema to the disease, and which are recorded both by Thomas and his collaborators and by Pittman as occurring in deciduous fruit trees.

Analyses of leaves of affected and normal plum trees showed that of the mineral constituents determined (Cl, N, SO_4, PO_4, Ca, Mg, K, Fe, Mn, Cu) the only ones which were present in lower amount in chlorotic than in normal plants were manganese and copper. Some of Anderssen's determinations are shown in Table 11. No effect was, however, produced by applying manganese to the chlorotic trees, but dipping chlorotic leaves in a solution of copper sulphate (0·3 p.p.m.) cured the chlorosis, and treatment of the soil with copper sulphate to the extent of 0·25–2 lb of copper sulphate per tree brought about recovery from the diseased condition. Isaac (1934) also reported the cure of chlorosis in young peach trees in South Africa by the application of copper sulphate to the soil, while manganese sulphate effected no improvement.

Die-back of apple trees ('wither tip' or 'summer die-back') as a result of copper deficiency has also been reported as occurring

7

TABLE 11. *Ash, manganese and copper content of leaves from chlorotic and normal Kelsey plum trees.* (Data from Anderssen)

Sample	Ash (per cent of dry weight)	Manganese (p.p.m. dry weight)	Copper (p.p.m. dry weight)
Top leaves chlorotic	10·56	10·0	3·2
Top leaves normal	8·85	17·1	7·3
Middle leaves chlorotic	12·34	14·7	2·9
Middle leaves normal	9·15	16·6	4·6
Lower leaves chlorotic	14·28	25·5	3·5
Lower leaves normal	10·83	19·5	6·9

in Western Australia by Dunne (1938), but the earlier symptoms of the disease do not seem to be quite the same as those described by Anderssen. Thus Dunne makes no reference to rosetting, but states that first brown spots and then small necrotic areas appear on the terminal leaves. Eventually the leaves wither and fall, and then follows the dying back of the shoot. Application of copper sulphate, either to the soil or by injection, was effective in bringing about arrest of the disease and recovery of the trees. The copper content of leaves from healthy trees varied from 5·5 to 12 p.p.m. of the dry weight, whereas in leaves from affected shoots the content varied from 1 to 3·6 p.p.m. of the dry weight. These values agree well with those found by Anderssen for the leaves of Kelsey plum trees, and by Oserkowsky and Thomas for the leaves of Bartlett pear.

The symptoms of copper deficiency in apples and pears, as they appeared on a plot at the Royal Horticultural Society's Gardens at Wisley and as recorded by Bould, Nicholas, Potter, Tolhurst and Wallace (1949), resembled those described by Dunne. There first appeared in early July large irregular necrotic areas in the terminal leaves which were followed by the upward curling and distortion of the leaves. The lower leaves of affected shoots tended to be pale green and to be affected with numerous small necrotic spots. Towards the end of July the leaves were shed from the apex of the shoots downwards and later in the summer the shoots withered and died.

Haas and Quayle made many determinations of the copper content of orange and lemon leaves and fruits, including samples from normal and exanthematic trees and from affected trees that had recovered as a result of the application of copper sulphate, and concluded that so much variation exists in the copper content of the leaves and fruits of trees from different localities and sites that it is not possible to decide from a knowledge of the copper content whether trees are suffering from a deficiency of that element or not. It may be said, however, that the values they obtained are of the same order as those already noted here. Thus the copper content of mature healthy orange leaves from untreated trees varied from about 7 to 15 p.p.m. of the dry weight, the corresponding values for the leaves of lemon being between 4 and 13 p.p.m. For two grapefruit leaves the values were 6·4 and 7·38 p.p.m. The fruit appears to contain much less copper calculated on a dry-weight basis, the content of the element in oranges and lemons ranging respectively from roughly 2 to 4 and 3 to 5 p.p.m. in these fruits.

Reclamation disease. A disease attributed to copper deficiency, which affects oats and other cereals, beet and leguminous crop plants, occurs on reclaimed heath and moorland soils in Denmark, Holland and other parts of Europe. In affected plants the tips of the leaves become chlorotic, and in cereals this is followed by a failure of the plants to set seed. This disease, known as reclamation disease or yellow-tip, was originally attributed to the toxic action of a constituent of peat, but Sjollema (1933) in Holland showed that the disease could be cured by the addition of copper sulphate to the soil, and that the content of copper in wheat, rye and hay grasses was raised by this soil treatment. Later, Gram (1936) obtained similar results with barley and oats in Denmark, and Undenäs (1937) with oats in Sweden, while in 1938 Piper found the disease affecting cereals in South Australia and also found that it could be controlled by the application of copper sulphate to the soil.

Oats were grown by Brandenberg (1933, 1934) in water culture in which the amount of copper present in the culture solution was carefully controlled. When copper was excluded, growth was poor and the plants developed the symptoms of reclamation disease. With small amounts of copper added to the solution vegetative growth was more normal, but fruiting was not unless the concentration of copper in the solution was at least 0·5 mg per litre. Later Piper (1942 a) carried out water-culture experiments with specially purified reagents in which the initial copper concentration of the culture solutions varied from 0 to 3 mg per litre. In addition to the ordinary major mineral nutrients the solutions contained small quantities of boron, zinc, manganese and molybdenum as well as sodium chloride. The main results were as follows. In the cultures without copper, 27 days after setting the seeds to germinate, growth and tillering were noticeably less than in cultures provided with copper. During the next fortnight these symptoms became more pronounced, the leaves were a paler green, the tips of the younger leaves appearing definitely chlorotic, withering and dying without unrolling, while the base of the leaf continued to emerge and grow. In subsequently emerging leaves these symptoms were more pronounced until growth of the tiller ceased and its death ensued.

In cultures provided with 3 μg of copper per litre definite symptoms of copper deficiency were not observed until 71 days after setting the seeds to germinate, although the plants made less growth than those with more copper. Slightly older leaves developed a marginal chlorosis. Ultimately the main tillers ceased growth and died, and although secondary tillers were produced these also developed the symptoms already described and ultimately stopped growing.

When cultures were initially provided with 6 μg of copper per litre more growth took place before the symptoms of deficiency developed in the youngest leaves. In these cultures the slightly older leaves exhibited a loss of turgor so that they appeared limp and drooping. Secondary tillers developed as before, but all stopped growing before ear production.

The plants in a culture solution containing initially 10 μg of copper per litre grew normally for 19 weeks before the symptoms of copper deficiency were evident. Flowering took place but grain was not formed. When the concentration of copper in the culture solution was 20 μg per litre some grain was produced but a proportion of the spikelets were sterile. In cultures grown in solutions containing from 50 to 500 μg of copper per litre growth was normal. If the concentration of copper was higher than this, a toxic effect was observed, growth and tillering being depressed while the leaves became stiff and erect. Although the leaves were first of a particularly deep green colour the younger leaves developed a strong chlorosis, which was more marked the higher the copper concentration of the culture solution.

The effect of copper concentration on the growth of oats is demonstrated by Piper's photograph of his cultures in Plate 9. The effects of copper deficiency on oat, broad bean and tomato are shown in Plates 10, 11 and 12, respectively.

Piper's work has confirmed the view that reclamation disease in oats is the result of copper deficiency. Similar experiments by Piper on other members of the Gramineae, namely, wheat, *Lolium subulatum* and *Phalaris tuberosa*, showed that these also are affected in the same way by shortage of copper.

5. The Symptoms of Molybdenum Deficiency

The recognition of molybdenum as a necessary constituent of many, if not all, plants came later than that of the four other generally accepted micro-nutrients. The symptoms ascribable to molybdenum deficiency displayed by tomato plants in water culture as first described by Arnon and Stout (1939b) are these. First the lower leaves develop a very characteristic mottling. This is followed later by necrosis at the leaf margins along with a characteristic curving over of the marginal regions of the leaf. Abscission of flowers takes place so that no fruit is produced.

The symptoms of molybdenum deficiency of oats grown in water culture were described soon after by Piper (1940). About the time of emergence of the panicles necrotic areas appear about midway along the lamina of the upper leaves, and these areas frequently extend right across the leaf. With a line of weakness so produced the leaf bends back sharply: the band is first smooth, but finally a kink develops and the middle necrotic region of the leaf dries out a light reddish brown. Although the inflorescence develops normally the grain consists entirely of empty husks.

So far only about four diseases common enough to have a distinctive name have been traced to deficiency of molybdenum. The first to be recognized was the condition known as whiptail which affects cauliflowers, savoy cabbages and other varieties and species of *Brassica*. The disease of *Citrus* called yellow spot, which was described by Floyd more than half a century ago, was traced to molybdenum deficiency by Stewart and Leonard in 1952. The scald disease of beans in Australia was attributed to shortage of molybdenum by Wilson in 1949 while the condition of *Hibiscus* called strap leaf also appears to result from molybdenum deficiency. Moreover it now seems certain that other species and varieties growing on natural and cultivated soils may be adversely affected by shortage of molybdenum. Soils deficient in molybdenum have been reported from various parts of the world including Great Britain and Holland, but particularly Australia, New Zealand and the United States where, according to Rubins (1956) molybdenum deficiency certainly occurs in thirteen states and possibly in at least three more. On such soils plants, particularly lucerne as well as cauliflowers and *Citrus*, may exhibit symptoms of molybdenum deficiency, while other crop plants, including lettuce, broccoli and clover, may also be noticeably affected. Hewitt (1956) has described the symptoms of molybdenum deficiency in more than forty species and varieties of higher plants, most of the information being derived from controlled culture experiments. The symptoms vary from species to species but chlorosis, either partial producing spotting or mottling, or general over the whole leaf, is usual. Wilting or scorching of the margins of the leaves follows and with the development of necrotic regions the leaves finally die (see Plate 13). According to Hewitt the symptoms of chlorosis, wilting and scorching arising in culture experiments are associated with the supply of nitrogen in the form of nitrate in the culture solution (cf. Plate 14).

Whiptail of cauliflower and other brassicas. The first description of the condition known as whiptail in the field was given by Ogilvie and Hickman in 1936; that it was a condition resulting from molybdenum deficiency was demonstrated by Hewitt and Jones (1947) who succeeded in reproducing the symptoms of whiptail as they appear in plants of *Brassica* varieties growing in the field by growing the plants in molybdenum-deficient sand cultures. The susceptibility of the different varieties are not the same, swedes and Brussels sprouts appearing to be the most readily affected. In such molybdenum-deficient plants pale chlorotic patches develop between the veins of the middle region of the stem and give the leaves a mottled appearance. The margins of the leaves turn grey-green and flaccid, then become brown, and finally the whole of the laminae wither so that the leaves consist of little more than the midribs, a few small irregular pieces of tissue being all that is left of the laminae.

Bean scald. A disease of dwarf beans (*Phaseolus vulgaris*) recorded only for the Gosford–Wyong district of New South Wales is characterized by an intervenal mottling followed by 'scald' or necrosis of both intervenal and marginal regions of the leaves. The disease was found to occur in plants grown from seed produced in plants grown on acid soils in the Gosford–Wyong area and sown in similar soil. Seed from plants grown in a number of areas outside the district produced plants free from the disease. The condition was attributed by Wilson (1949) to deficiency of molybdenum. Application of a solution of sodium molybdate to affected plants resulted in the intervenal areas developing a healthy green colour within a few days, while there was a reduction of the high concentration of oxidizing substances, probably mainly nitrates, present in the chlorotic areas. Supporting evidence was provided by the fact that whiptail of cauliflowers occurred in the same areas as bean scald while the tendency for scald to develop was increased by adding sulphur to the soil which rendered it more acid, and reduced by application of lime which rendered it more alkaline, the availability of molybdenum increasing with increase in pH of the soil (see p. 140). That plants grown from seed obtained from outside the Gosford–Wyong district did not develop scald was attributed to this seed containing a sufficiency of molybdenum.

Yellow spot of *Citrus*. The production of yellow chlorotic spots in the leaves of *Citrus* was described by Floyd in 1908. Stewart and Leonard recorded its incidence on acid soils in Florida and noted that it affected temple orange, sweet orange, tangerine and grapefruit, trees on grapefruit rootstock being the most susceptible. Yellow spot has indeed been recorded as occurring in all varieties of *Citrus* throughout the whole of the *Citrus* growing areas of Florida. The symptoms of yellow spot in the *Citrus* varieties examined by Stewart and Leonard were described by them as follows. In the leaves of the early summer flush, areas with a water-soaked appearance develop and these later become large, round or oblong, and yellow. These yellow spots may occur anywhere in the leaf, sometimes irregularly, sometimes in regular sequence between the main veins, but most frequently inside the margin. The spots may coalesce, producing large affected areas and the leaves may finally fall so that the tree becomes badly defoliated. The symptoms of molybdenum deficiency as these develop in molybdenum-deficient water cultures were described by Vanselow and Datta (1949) as a mottling of the leaves followed by necrosis.

Strap leaf of *Hibiscus*. In the condition known as strap leaf reported as occurring in *Hibiscus* affected leaves are described as dark green but stunted and constricted laterally, while they exhibit prominent and distorted parallel veins. In the rozelle (*Hibiscus sabdariffa*) molybdenum deficiency has been found by Von Stieglitz and Chippendale (see Stout and Johnson, 1956) to bring about collapse of the massive central portion of the flower. This is of some economic importance as the fruit is used for making a jelly.

Molybdenum deficiency in lucerne. Lucerne or alfalfa (*Medicago sativa*) is very sensitive to molybdenum deficiency (Plate 14). According to Reisenauer (1956) molybdenum-deficient lucerne plants in the field appear stunted and pale green, the leaves having intervenal chlorotic spots which spread and coalesce until they affect the whole leaf which then dies and falls. This particularly concerns the lower leaves and with severely affected plants only the upper leaves remain. Similar symptoms were reported by Evans, Purvis and Bear (1950) as occurring in plants grown in molybdenum-deficient cultures; in addition they reported a marginal necrosis of the upper leaves. The symptoms of plants growing in the field were considered by Reisenauer to be those of nitrogen starvation and this conclusion was supported by analysis of plants showing varying degrees of molybdenum deficiency. As Table 12 shows, the severity of the symptoms is related rather more to the nitrogen content than to the molybdenum content of the leaves.

TABLE 12. *Molybdenum and nitrogen contents of the leaves of molybdenum-deficient plants of lucerne*

Symptoms	Nitrogen content of leaves (per cent)	Molybdenum content of leaves (p.p.m.)
None	3·61	0·34
Mild	2·98	0·26
Moderate	2·68	0·26
Severe	2·63	0·28

6. THE SYMPTOMS OF CHLORINE DEFICIENCY

Chlorine has not been generally considered an essential element for plant growth but from time to time investigators have concluded from experimental observations that chlorine was necessary for the growth of one plant or another. It has already been mentioned that about a century ago Nobbe and Siegert regarded chlorine as essential for the growth of buckwheat, a finding for which confirmation was obtained by the work of Lipman in 1938. In the meantime Mazé in 1915 had decided that chlorine was essential for the growth of maize. More recently Eaton (1942) reported increased growth of tomatoes and cotton as a result of the application of fertilizers containing chloride, and Raleigh (1948) found increased growth of garden beet in water cultures as a result of the addition of chloride to the culture solution. More recently still evidence of the essentiality of chlorine for tomato was presented by Broyer, Carlton, Johnson and Stout (1954) and for sugar beet by Ulrich and Ohki (1956) who, however, thought that chlorine could be replaced by bromine. Johnson, Stout, Broyer and Carlton (1957) were able to induce acute chlorine deficiency in lettuce, cabbage, barley, lucerne and beans, as well as in several of the species noted above, namely, tomato, sugar beet, buckwheat and maize. Squash was an exception but it did not follow that this plant did not require chlorine for, as well as maize and beans, it was able to acquire a supply of chlorine from the atmosphere. Of all the plants examined lettuce was the most susceptible to injury resulting from chlorine deficiency.

Symptoms of chlorine deficiency have been described by both Broyer and his co-workers and by Ulrich and Ohki. In tomato, wilting of the tips of the leaflets occurs, there is a progressive chlorosis of the leaves followed by bronzing and necrosis, and in severe cases fruit formation fails. In sugar beet an intervenal chlorosis develops in the middle leaves of the plant which at first resembles the effect produced by manganese deficiency, the leaves having a mottled or mosaic appearance at first visible only by transmitted light. These intervenal areas later appear as light green to yellow smooth and flat depressions. Secondary roots are also affected, having a stumpy appearance.

CHAPTER IV

THE EFFECTS ON PLANTS OF TRACE-ELEMENT EXCESS

WHILE the trace elements are necessary for the normal growth and development of plants, excess of them may be as injurious as their deficiency. Trace elements in excess are, in fact, poisons, and toxic symptoms characteristic of one or other of the trace elements in their action on particular species have been recorded in a number of instances. Such injurious effects are, of course, not limited to micro-nutrients and may be produced by a large number of elements, particularly the so-called heavy metals. The amount of any element which has to be absorbed in order to produce injury varies not only with the element and the plant but also probably with other factors such as the age of the plant or organ and the amount of other elements present in the medium and the plant.

MANGANESE

The visual effect of excess of manganese on the cereals wheat and barley is the browning of the roots and the subsequent formation of brown spots in the leaves. This was shown as long ago as 1902 by Aso, while more recently this symptom of manganese excess in barley leaves was recorded by Williams and Vlamis (1957). More commonly excess of manganese has been found to produce chlorosis. This was reported in pineapple plants by Kelley (1909, 1912), in oats by Rippel (1923), in *Citrus* by Haas (1933) and in dwarf bean (*Phaseolus vulgaris*) and common vetch (*Vicia sativa*) by Löhnis (1950). In the dwarf bean excess of manganese induces chlorosis which at first affects the leaf margins and then in the younger leaves extends to the areas between the veins. As the leaves age they become crinkled while in the intervenal regions there appear whitish blotches and finally necrotic spots. The petioles and older parts of the stem are also affected, brownish purple spots developing on them. In the vetch the intervenal chlorosis is accompanied by a purplish discoloration of the margins of the leaflets.

The development of purplish spots on the primary leaves as a symptom of manganese toxicity was observed in cowpeas by Fujimoto and Sherman (1948). Later these spots became brown and necrotic and the leaves died. The first-formed trifoliate leaves were very chlorotic and on these also brown necrotic spots developed which then extended into the surrounding areas and then these leaves also died. Morris and Pierre (1949) also observed the formation of small reddish purple spots on the leaves of the cowpea as a result of excess manganese, the number of spots increasing with increase of the concentration of the manganese supplied in the medium. The leaf margins were not chlorotic. These later workers also described the symptoms of manganese toxicity in some other members of the Leguminosae. In *Lespedesa* the symptoms consisted in the formation of dark reddish brown spots which in severely affected plants accounted for half the total leaf area. In these the leaves exhibited a marginal chlorosis affecting up to 75 per cent of the margin, only the region near the base of the leaf not being affected. In soya beans excess of manganese resulted in the appearance of pale green irregular intervenal areas, while when supplied with manganese in higher concentrations many leaves became brown and freshly formed leaves appeared crumpled. As with cowpeas no marginal chlorosis was observed. The crinkling of soya bean leaves along the margins was noted by Evans, Lathwell and Mederski (1950) as a symptom of manganese toxicity, but they found also that excess manganese produced a marked necrosis of the leaf margins while the rest of the leaf remained dark green.

The effects of excess manganese on sweet clover were found by Morris and Pierre (1949) to be in some contrast to those on cowpeas for in the former no spotting of the leaf was seen but there was marked chlorosis of the leaf margin and wrinkling of the leaf. Plate 16 A shows the effects on cauliflower: purple-brown necrotic spots on leaf-margins, intervenal brown necrotic mottling and cupping of the leaf-margins.

Different species appear to vary very much in regard to their susceptibility to injury in consequence of excess manganese. Aso (1902) found peas less susceptible than wheat and barley, while Löhnis (1950) found no toxic symptoms in cauliflower, strawberry, potato, tobacco and oats subjected to the same conditions of high manganese supply as those producing injury in dwarf beans. Olsen (1934, 1946) found a wide range in sensitivity to manganese in five species he examined. Thus a culture solution containing 0·5 p.p.m. of manganese proved toxic to *Lemna polyrrhiza* and *Senecio sylvatica* while a concentration as high as 62·5 p.p.m. was necessary to produce injury in maize.

Morris and Pierre (1949) examined the effect of the concentration of manganese in the nutrient solution on plants of five different species of Leguminosae. Even within the same family notable differences in sensitivity to manganese were observed with different species. Some of their results which bring out this are shown in Table 13. The values of dry weights are for the total number of plants grown in a culture vessel which was constant for each species, namely, six for sweet clover and three for each of the other species. With soya beans and cowpeas the numbers shown in the table are for shoots only; the roots of these plants appeared not to be affected by higher concentrations of manganese.

TABLE 13. *Effect of manganese concentration in the nutrient solution on the growth of different plants.* (Data from Morris and Pierre)

Concentration of manganese in solution in p.p.m.	Dry weight of plants in grams				
	Lespedesa	Soya bean	Cowpea	Peanut	Sweet clover
0·1	1·09	6·84	6·97	7·3	5·82
1·0	0·81	6·17	6·10	6·93	—
2·5	0·63	6·89	6·56	7·01	6·45
5·0	0·51	6·35	6·11	6·91	4·80
10·0	0·42	4·13	4·72	5·56	4·19

It will be observed that with *Lespedesa* there was a significant decrease in growth resulting from the presence of 1 p.p.m. of manganese, as compared with 0·1 p.p.m., whereas with peanut a definite effect on growth was only observable with a concentration of 10 p.p.m. Even in this strongest solution there were only slight visual symptoms of manganese toxicity, a few of the leaves exhibiting a slight marginal chlorosis.

Analyses of plants exhibiting manganese injury showed that these contained a higher proportion of manganese than normal plants. Thus in one year Löhnis found dwarf bean plants containing 1210 p.p.m. of manganese or more exhibited toxic symptoms, while in another year this was so with plants containing 1285 p.p.m. or more of manganese. These amounts compared with 536 p.p.m. or less of manganese in normal plants. Where plants are less susceptible to injury this may be due either to a greater tolerance or to a lower uptake of manganese. Potato appears to be an example of the former, for Löhnis found plants of this species with a high content of manganese exhibiting no sign of injury, while strawberries, and oats to an even more pronounced extent, were found to possess a lower proportion of manganese than dwarf beans growing under similar conditions. There appears, indeed, to be a great divergence among different species in regard to their absorption of manganese. In experiments with twenty different species of flowering plants Collander (1941) found that some had a capacity for absorbing manganese twenty to sixty times that of species with a minimum capacity for absorbing this element.

ZINC

Records of the harmful effect of excess absorption of zinc go back for almost a century. Early descriptions of the effect of excess of zinc on barley by Krausch and Storp in 1882 and 1883 respectively indicated that the leaves of affected plants became chlorotic and developed small reddish brown intervenal patches. Other grasses examined, including timothy, although slightly more resistant, behaved similarly. With willow also excess of zinc was found to produce chlorosis, while it was noted that the roots turned a brownish colour.

The symptoms of zinc toxicity in soya beans were described by Earley in 1943. At first a red pigment was observed to form at the base of the middle vein of the leaf and next the leaves began to curl under. Chlorosis of the tripartite leaves towards the upper part of the stem now appeared, the stem apex died and the red pigment became more intense in the leaf veins and also in the leaf petioles and the stem. More recently Hewitt (1953) observed that with sugar beet grown in sand cultures containing an excess of zinc, symptoms resembling those of manganese deficiency developed.

As with manganese, different species exhibit considerable differences as regards their susceptibility to zinc poisoning. In

1885 Baumann recorded the time taken by plants of a number of species to die when grown in a solution containing 44 mg of zinc sulphate per litre. These times ranged from 16 days with *Trifolium pratense* to 194 days with *Onobrychis sativa*. Barley plants were killed in 30 days and beet in 76 days.

Even varieties of the same species may differ significantly in their sensitivity to excess of zinc. Thus of two varieties of soya bean Earley found one, Peking, tolerated a concentration of 0·1 p.p.m. of zinc but was damaged by a concentration of 0·2 p.p.m. in the medium, and killed by one of 0·4 p.p.m. Plants of the other variety, Hudson Manchu, tolerated zinc in a concentration of 0·8 p.p.m. and were killed by a concentration of 1·6 p.p.m.

BORON

The toxic effect of boron in excess on the growth of plants appears first to have been noted by Knop in 1884 and has been frequently confirmed subsequently. As with other elements different species show considerable differences in their sensitivity to injury by boron. Thus Brenchley (1914) recorded that in experiments carried out at Rothamsted young barley plants were injured by boric acid in a concentration of 4×10^{-6} while a concentration of not less than 2×10^{-5} was necessary to injure young garden pea plants. Earlier Hotter (1890) had found that garden pea plants were injured in nutrient solutions containing boric acid in half this concentration. Brenchley found maize less sensitive to boron excess than peas and oats.

According to Chapman and Vanselow (1956) the concentration of boron in some irrigation waters may be sufficiently high to exert a toxic effect on *Citrus* trees. They concluded that if the concentration of boron in these waters exceeded 1 p.p.m. a harmful effect of boron was evident while double this concentration rendered the conditions unsuitable for the growth of *Citrus*.

The chief visual symptom of boron poisoning in peas as described by Hotter is the appearance of brown areas in the lower leaves, particularly at the margins; later these extend to the upper leaves. Brenchley recorded that in peas exhibiting boron poisoning the tip and margins of the leaves are first affected and the discoloration then extends inwards without the formation of isolated brown spots. In barley, however, she noted that the leaves turned yellow with large brown spots so that the leaves appeared mottled. Williams and Vlamis (1957) also noted the formation of large blotchy necrotic patches in barley leaves as a result of excess of boron and contrasted these with the small spots produced by manganese poisoning. The effect of excess boron on lucerne, according to Wallace and Bear (1949), is a marginal brown necrosis of the lower leaves and would thus appear to be similar to that found by Brenchley in the garden pea, another member of the Leguminosae.

According to Evans, Lathwell and Mederski (1950) the first sign of a toxic effect of excess boron observed in soya beans is the necrosis of the margins of the primary leaves. Later the tripartite leaves of the lower and middle parts of the stem become necrotic. Finally the youngest leaves become yellow, crinkled and necrotic. Evans and his co-workers found that soya beans were particularly sensitive to boron, a concentration of 0·5 p.p.m. in the nutrient solution being sufficient to produce slight symptoms of boron toxicity, while with a concentration of 2 p.p.m. severe toxicity symptoms developed a week after the emergence of the seedling from the seed. Beeson, Gray and Hamner (1948) also noted the sensitivity of soya beans to anything but low concentrations of boron.

In sand cultures of hops Cripps (1956) found symptoms of boron toxicity developed when the medium contained 10 p.p.m. of boron in presence of other nutrients, and this was so whether calcium, which had been found to influence uptake of boron (see p. 134) was supplied at any of the three levels of 20, 100 and 500 p.p.m. The toxic symptoms consisted of chlorosis of the leaves, most noticeable at the margins, a downward curling of the leaves and poor leaf lobation. There followed a formation of brown spots about 1 mm in diameter in the intervenal areas particularly near the margins, which became scorched with a tendency to curve upwards. The intervenal areas became more and more chlorotic and the scorching extended inwards from the margins and the leaves so affected eventually died. Inflorescences were few and small, especially in the plants receiving calcium at the highest level.

McIlrath and de Bruyn (1956) described the symptoms of boron toxicity which developed in sand cultures of Siberian millet

(*Setaria italica*) as a burning of the leaf tips followed by the appearance of necrotic spots along the margins of the leaves. These symptoms were observed in plants supplied with boron at the levels of 5 and 50 p.p.m. In the higher concentration the necrosis increased with age and the lower leaves finally died. Along with these symptoms there generally resulted a decrease in both total and percentage dry weight and an increase in the percentage of ash in the shoots.

Chapman and Vanselow found that moderate symptoms of boron toxicity were shown by leaves of *Citrus* containing 200 p.p.m. of boron in the dry matter.

COPPER

That copper in any but very low concentration has a toxic action on plants has been known for a very long time. Nearly seventy years ago it was found by Otto that ordinary distilled water obtained from a copper still contained a high enough concentration of this element to induce a reduction in the rate of growth of garden pea, dwarf bean and wheat grown in culture solutions made up from this water. The experiments of Piper with barley demonstrated very clearly the effect of increasing copper concentration above 0·5 p.p.m. in bringing about reduction in the rate of growth of barley (cf. Plate 9).

The toxicity of copper, like that of other poisons, can be greatly affected by the presence of other substances. For example, Brenchley found the growth of barley was stopped in a solution of copper sulphate alone in a concentration of 1 p.p.m. but not in a solution containing copper sulphate in a concentration of 4 p.p.m. together with nutrient salts.

Most of the accounts of the toxic effects of copper have dealt only with the decrease in growth and final death resulting from excess absorption of this element, and few observations are on record on the effects of copper on physiological processes or of visual symptoms of injury produced by excess of copper. Treboux (1903), however, recorded a reduction of 20 to 25 per cent in the rate of photosynthesis of shoots of *Elodea* when these were transferred from water to a solution of copper sulphate in a concentration of 10^{-6}N, the initial rate being only partially restored on re-transference to water, thus suggesting some

PLATE 1

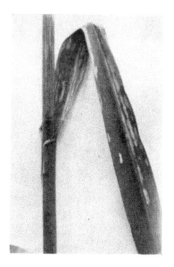

A. Oat seedling affected with grey speck. Note the grey areas and the characteristic line of weakness in affected leaves.

B. Oat leaf exhibiting the characteristic symptoms of grey speck.

C. Sugar beet suffering from manganese deficiency. Note the speckling of the leaves, appearing in the photograph as a mottling.

PLATE 2

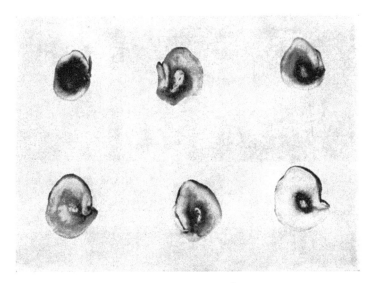

Peas suffering from marsh spot.

PLATE 3

Zinc deficiency in oat. Note the white necrotic lesions appearing
first in the lower leaves.

PLATE 4

Zinc deficiency in broad bean. Note the progressive decrease in size attributable to exhaustion of the zinc reserve in the seed as the plant develops. The leaflets become narrower and the internodes become shorter.

PLATE 5

A. Sugar beet suffering from boron deficiency. Note the stunted and curled appearance of the young inner leaves.

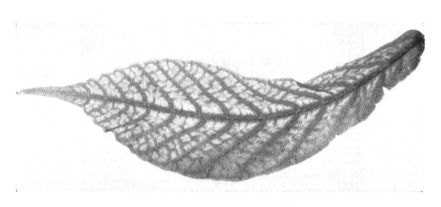

B. Leaf of cacao suffering from zinc deficiency.

PLATE 6

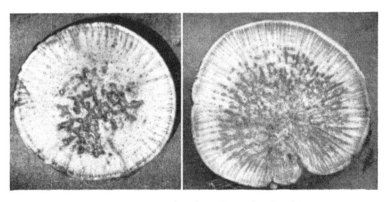

A. Transverse section through swedes showing
characteristic appearance of brown heart.

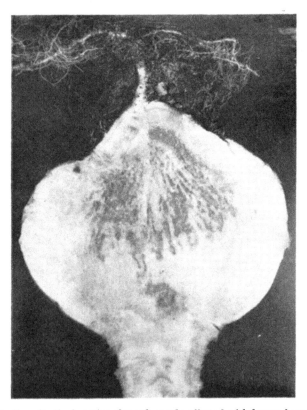

B. Longitudinal section through swede affected with brown heart.

PLATE 7

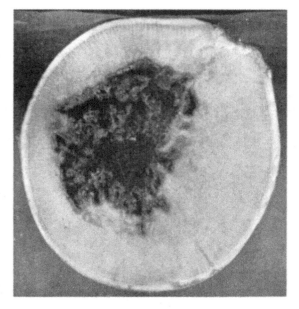

A. Late stage in boron-deficient swede.

B. Boron-deficient cauliflower. Note patches of discoloration of head.

PLATE 8

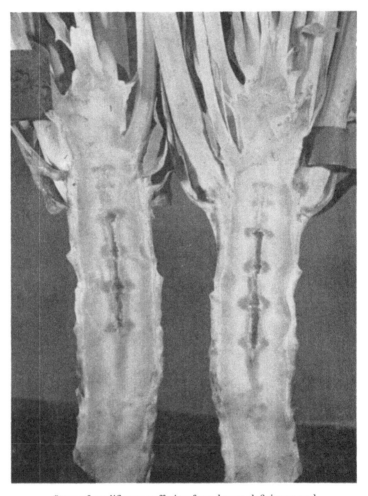

Stem of cauliflower suffering from boron deficiency and
showing very typical breakdown of central tissue.

PLATE 9

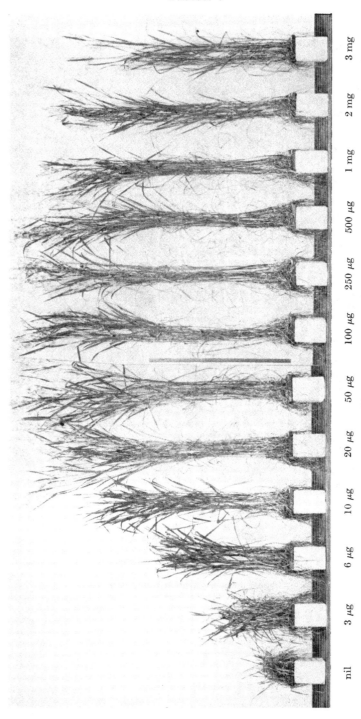

Effect of concentration of copper on the growth of oats. The corresponding copper values are expressed as weights per gram of nutrient solution.

nil   3 µg   6 µg   10 µg   20 µg   50 µg   100 µg   250 µg   500 µg   1 mg   2 mg   3 mg

PLATE 10

Copper deficiency (wither tip) in oat.

PLATE 11

Copper deficiency in broad bean. The plant on the left exhibiting reduction in size of the apical leaves, rolled and necrotic at the tips, and with flowering suppressed, was grown from seed deficient in copper; the plant on the right was grown from normal seed.

PLATE 12

Copper deficiency in tomato. Leaves from copper-deficient plants
are shown above, a normal leaf below.

PLATE 13

Tomato plants suffering from severe molybdenum deficiency.

PLATE 14

Molybdenum deficiency in lucerne. The plants received
a supply of nitrogen as nitrate.

PLATE 15

Cauliflower plant grown in absence of molybdenum.

PLATE 16

A. Cauliflower plant suffering from manganese toxicit

B. Sugar beet showing symptoms of iron deficiency
induced by excess of copper.

permanent injury to the photosynthetic mechanism. As regards visual symptoms of copper toxicity, Hewitt (1953) found that excess copper led to chlorosis of the leaves of tomato and sugar beet resembling that produced by a shortage of iron (Plate 16 B), but oats and kale appeared not to be affected in this way.

MOLYBDENUM

The question of excess absorption of molybdenum is economically important in agriculture as relatively large amounts of this element in herbage plants may lead to very serious disease in cattle, a matter which is dealt with in a later chapter.

The effect of molybdenum absorbed in sufficient quantity to be toxic was examined in several species by Warington. Three members of the Solanaceae, namely, potato, tomato and *Solanum nodiflorum* were found to be much more sensitive to excessive amounts of molybdenum than broad bean and barley. The usual symptom observed was the production of a distinctive chlorosis. In the potato the shoots became reddish yellow and in tomato and *Solanum nodiflorum* a golden yellow. Millikan also observed a golden yellow chlorosis in flax resulting from excess molybdenum if nitrogen was supplied as a nitrate, including ammonium nitrate, or urea, but not if the nitrogen was supplied as an ammonium salt other than ammonium nitrate. This distinctive chlorosis appears to result from the formation of a molybdenum-tannin compound present in the cells in the form of yellow globules. Where anthocyanins are present excess molybdenum results in the formation of blue granules, presumably of a molybdenum-anthocyanin compound.

Different plant species vary greatly in their capacity for absorbing molybdenum. This is of considerable importance in regard to the maintenance of cattle on land containing a relatively high content of molybdenum as the floral composition of the pastures will determine their molybdenum content and so the amount of this element taken into stock feeding on the pastures. This question is dealt with in chapter VII.

CHAPTER V

FACTORS INFLUENCING
THE ABSORPTION OF TRACE ELEMENTS
AND THEIR EFFECTS ON PLANTS

REFERENCES have been made in the course of the preceding pages to researches in which it has been found that, with concentrations of one or other of the trace elements below a certain minimum, symptoms of deficiency of that element occur. Similarly, as regards the toxic effects of excess concentrations, limits of concentration have been indicated above which injury results. It is, however, clear that minimum and maximum values of this kind have no very precise significance, for the rate of absorption of a mineral constituent depends on a number of variable factors of which the concentration of the element is only one. Other factors may include, first, the presence of other substances and ions in the medium and their nature and concentration, and, secondly, conditions in the plant which are largely a matter of speculation but the effect of which is indicated by the different absorptive capacities of different plants. Where the medium from which the trace element is absorbed is soil, the content of the element in the soil may be little guide to its actual concentration in the soil solution, the medium from which the element is directly absorbed. Here the nature of the compound containing the element, other substances present in the soil, both mineral and organic, the hydrogen-ion concentration and the activity of micro-organisms, factors which are by no means all independent of one another, may all play a part in determining the actual amount or concentration of the element available for absorption.

Data have also been presented indicating that plants suffering from a deficiency of an element have a lower content of it than normal healthy plants, while plants exhibiting injury through excess of it contain more of the element than normal plants. While these differences may be usual they are not universal, as examples given earlier indicate, and the content of an element in a plant is thus not an absolute criterion of whether it is present in deficient or excess quantity. The state in which the element is held in the plant would thus appear to be of importance in determining its effect on growth. This has been particularly brought out in work with manganese.

MANGANESE

Of the manganese contained in the soil only a relatively minute proportion is likely to be dissolved in the soil water, and so immediately available to the plant, but as this manganese is absorbed it will be replaced in the soil solution from exchangeable manganese held by adsorption to colloidal particles in the soil. A variety of solid manganese compounds may occur in soils, the manganese being divalent, tetravalent, and probably also trivalent. The divalent manganese is soluble and so forms a reserve from which the exchangeable and available manganese can be replenished. How far the practically insoluble trivalent and tetravalent forms contribute to available manganese is doubtful and has been disputed, but it is reasonable to conclude that the divalent manganese is the main source of supply of this element to the plant. However, oxidation or reduction of the soil manganese means changes in its valency so that the quantity which can be a source of manganese available for the plant can alter. Alkalinity and aeration further oxidation, that is, the formation of trivalent and tetravalent manganese from divalent, while acidity and shortage of oxygen lead to the reverse transformation. A lower hydrogen-ion concentration of the soil thus tends to reduce the content of manganese which can become available to the plant. The action of soil micro-organisms is in the same direction as they compete with higher plants for the manganese available and bring about its oxidation. According to Quastel the activity of soil micro-organisms in effecting the oxidation of manganese is at a maximum when the pH of the soil is from 6·0 to 7·9, that is, at about neutrality.

The well-known effect of over-liming of soil in inducing symptoms of manganese deficiency is usually attributed to its lowering the hydrogen-ion concentration and so reducing the amount of soluble and available manganese. Heintze (1946), for example, found the amount of exchangeable manganese fell steadily with addition of lime which raised the pH. Lime was added to an acid soil with a pH of 4·5 in amounts of 1, 3, 8 and

12 tons per acre which had the respective effects of raising the pH to 5·5, 6·3, 7·5 and 7·9. The exchangeable manganese content of the original soil was 10·2 mg per cent; with the progressively increasing additions of lime indicated it fell to 10·0, 8·4, 5·0 and 2·0 mg per cent.

The matter may, however, be not so simple as this would suggest, for Lundegårdh came to the conclusion that symptoms of manganese deficiency in the form of grey speck developed in oats growing on neutral and slightly alkaline soils but not on strongly alkaline as well as strongly acid soils, while Maschhaupt (1934) and Popp et. al (1934) found that a heavy dressing of calcium oxide would suppress the development of grey speck although the calcium oxide could not be replaced by calcium carbonate in this respect. A similar effect was recorded by Gisiger (1950) with an acid peat soil on which grey speck developed in oats when the soil was given a moderate dressing of calcium oxide but not when the dressing was heavy. The appearance of grey speck in this instance was correlated with the amount of manganese absorbed by the oat plants, the amount at first decreasing with increase in the amount of added lime and then increasing with further increase in the amount of calcium oxide added to the soil. Gisiger also examined the combined effect of additions of potassium hydroxide and calcium oxide to the soil on the incidence of grey speck and the yield of grain in oats and found that the development of grey speck induced by the addition of potassium hydroxide could be prevented by the addition of calcium oxide. Similarly, while the yield of grain fell markedly with increase in the amount of potassium hydroxide added to the soil, this effect could be reduced or eliminated altogether by addition of calcium oxide.

Gisiger's general conclusion was that the development of grey speck was related to the hydroxyl-ion concentration of the soil, this development first increasing with increase in the hydroxyl-ion concentration and then falling with further increase in pH. It is tempting to relate this finding with the activity of micro-organisms where the maximum oxidative activity was found to be round about neutrality.

The amount of manganese in soils available to plants may also be influenced by the concentration of iron salts. Boken (1955,

1956, 1957) found that the uptake of manganese by plants growing on soils containing trivalent and tetravalent manganese was increased as a result of the addition of ferrous sulphate to these soils. This could be explained as due to the ferrous salt bringing about the reduction of the manganese to the divalent condition in which it would be made more readily available to plants, as we have seen above.

It has been observed that symptoms of manganese deficiency are liable to occur in plants growing on soils containing much organic matter, such as old garden soils rich in humus. In some instances, as noted below, this is due to an actual low content of manganese in the soil, but otherwise it is probably due to the combination of manganese with constituents of the humus to form complex compounds in which the manganese is non-available.

Many determinations of the manganese content of soils have been made. In particular reference may be made to the very large number of semi-quantitative analyses of natural soils in Finland made by Lounamaa (1956). These soils included those derived from siliceous, ultrabasic and calcareous rocks and from habitats where the content of some particular element was known to be exceptionally high. In siliceous soils the manganese content ranged from 30 to 6000 p.p.m., in soils derived from ultrabasic rocks from 300 to 3000 p.p.m. and in calcareous soils from 300 to 6000 p.p.m. These are wide ranges and although values outside them have been registered, most determinations of the manganese contents of soils by other investigators fall within these limits. Thus the manganese contents of a number of soils of West Pakistan analysed by Wahhab and Bhatti (1958) ranged from 370 to 734 p.p.m. Particularly low contents of manganese have been found in soils containing a high content of organic matter such as peat and bog soils, while the manganese content of two podsolic soils in the United States were found by Connor, Schimp and Tedrow (1957) to be 20 p.p.m. of the oven dry weight as compared with 121 and 106 p.p.m. for two calcareous soils, and 460 and 520 p.p.m. for two soils formed on glacial drift mainly derived from a local acid red shale.

Few data exist of the concentration of manganese in the actual soil solution but no doubt it is usually very low. Determinations made by McGeorge (1924) for a number of soils in Hawaii were nil for soils with pH of 6·0, 7·76 and 7·9 although one of these soils, that with a pH of 7·76, had a high manganese content. For two soils with a pH of 4·46 values of 0·58 and 13·3 p.p.m. were obtained while for soils with pH of 4·93 and 5·39 the respective values of 2·9 and 2·4 p.p.m. were found. These data indicate that pH may be an important, but not the only, factor in determining the concentration of manganese in the soil solution.

Many estimates of the available manganese have been made by determining the exchangeable manganese, that which is extracted from the soil by a solution of some salt such as ammonium acetate or potassium nitrate. Such determinations can give only a rough idea of the manganese available to plants; for example, the amount of manganese replaced by the cation of the extractive depends on the cation. The values of exchangeable manganese so obtained exhibit almost as wide a range as that of the manganese content of soils and certainly indicate that soils vary considerably in the amount of manganese they contain which is available for plants.

This being so it is to be expected that plants will exhibit considerable differences in their manganese content in relation to the differences in the amount of available manganese in the soil and numerous analyses show this is so. For example, Hunter (1953) found the manganese content of the dry matter of fronds of bracken (*Pteridium aquilinum*) from some Scottish soils ranged from 32 p.p.m. on a serpentine soil to 484 p.p.m. on a granitic soil. The great number of semi-quantitative determinations of trace elements in plants growing on soils in Finland made by Lounamaa provide very many instances of the difference in manganese content of plants of the same species growing on different soils. To take two examples from about 150 species examined, the manganese content of the ash of stems of sheep sorrel (*Rumex acetosella*) ranged from 300 to 10000 p.p.m., while that of the leaves of the rose-bay willow herb (*Chamaenerion angustifolium*) ranged from 600 to 30000 p.p.m.

Different species growing in the same habitat may also exhibit wide differences in their absorption of manganese. Thus Thomas and Trinder (1947) found the manganese content of the ash of plants of *Molinia coerulea*, *Scirpus caespitosus* and *Vaccinium myrtillus* gathered from the same habitat in Northumberland on the same day (August 15) to be respectively 330, 502 and 2639 p.p.m. The manganese content of a number of grasses growing on the same soil, a fine sandy loam, was found by Beeson, Gray and Adams (1947) to cover quite a wide range, from 815 p.p.m. in red top to 96 p.p.m. in para grass.

There is some evidence that the content of phosphate in the soil may affect the availability of manganese. In a study of the effect of phosphate on the trace element nutrition of young plants of sour orange Bingham, Martin and Chastain (1958) found that additions of calcium phosphate to the soil enhanced the uptake of manganese by the plants. Thus additions of calcium hydrogen phosphate to a sandy loam had the effect of increasing the manganese concentration in the leaves from 21 p.p.m. in plants growing on soil without added phosphate to 28, 37 and 67 p.p.m. with dressings of 360, 900 and 1800 lb of phosphorus as calcium hydrogen phosphate per acre respectively. It was suggested that this increase in manganese absorption might be due to the increase in soluble manganese in the soil through the production of soluble manganese phosphates.

The addition of flowers of sulphur to soil, by reducing its alkalinity, may render more manganese available for absorption by plants. This was demonstrated by Owen and Massey (1953) who observed in roses grown in glasshouses certain symptoms which appeared to be induced by manganese deficiency. These included intervenal chlorosis of the leaves, reduction in the size of the leaves and weakening of the flower-bearing branches. The manganese content of affected leaves was less than that of normal leaves. In a variety 'Richmond', the manganese content of leaves displaying these symptoms was 23·5 p.p.m. but in plants growing in soil treated with flowers of sulphur it was 106·5 p.p.m.

There is abundant evidence that the absorption of manganese is influenced by the presence of a number of other substances in the medium. Prominent among these are iron salts. That the presence of iron brings about a reduction in the absorption of manganese is indicated by the finding by a number of investigators that the symptoms of manganese toxicity are reduced or eliminated altogether by the presence of iron. This effect was demonstrated in peas, radish, wheat and barley by Aso in 1902,

since when it has been shown in a number of other plants. Several
writers have urged that growth is at a maximum when manganese and iron are present in a certain ratio. So Pugliese (1913)
considered the optimum ratio of manganese to iron in the culture
solution for the growth of wheat was 1 to 2·5. Tottingham and
Beck (1916) wrote of an antagonism between iron and manganese
in the growth of wheat while Scharrer and Schropp (1934) found
that with maize in water culture the growth of the roots was at a
maximum when the ratio of manganese to iron in the culture
solution was 1 to 7.

The results of an elaborate series of culture experiments with
soya bean described by Somers and Shive (1942) emphasized the
importance of the iron/manganese ratio. In these experiments
eighteen different combinations of iron and manganese were used
in the culture solutions. The iron content varied from 0·005 to
3·00 p.p.m. and the manganese content from 0 to 5·00 p.p.m.;
the various combinations are shown in Table 14. Actually complete absence of manganese was never attained in the culture
itself since there was some present in the seed used and no doubt
a little would also be introduced as impurity either from culture
vessels or other nutrient salts used. The culture solutions also
contained the usual major nutrients and boron. The concentrations of iron and manganese and a pH of 4·6–4·8 (to prevent
precipitation of iron) were maintained approximately constant by
the use of a technique in which a continuous flow of solution
passed through the culture vessels, and the solutions were completely changed every other day.

Approximate determinations of both soluble and insoluble
iron and manganese in the tissues were also made. To separate
the soluble and insoluble fractions the fresh material was frozen
by means of an ice–salt mixture and from this, in thawing, the
juice was expressed under a pressure of 1600 lb per sq. in. applied
for 2½ min. The expressed juice together with washings of the
press cake and muslin containing it were taken as containing the
soluble iron and manganese, the press cake the insoluble fraction.

A study of Table 14 shows at once how the dry weight and
condition of the plants is related to the Fe/Mn ratio and not to
the absolute concentrations of these nutrients. Thus plants
growing in solutions containing 0·002, 0·250 and 2·00 p.p.m.

TABLE 14. *Effects on soya beans grown in water culture of different proportions of iron and manganese.* (From Somers and Shive)

Concentration in substrate (p.p.m.)		Ratio Fe/Mn in substrate	Ratio soluble Fe/soluble Mn		Dry weight per plant (g)	Condition of plants
Fe	Mn		In leaves	In roots		
0·005	0·000	—	2·65	5·04	3·47	Normal
0·005	0·002	2·50	2·37	3·36	3·78	Normal
0·005	0·010	0·50	2·05	2·73	3·62	Slight Fe deficiency (Mn excess)
0·005	0·250	0·02	1·00	1·44	2·86	Medium Fe deficiency (Mn excess)
0·005	2·00	0·0025	0·80	1·35	1·50	Severe Fe deficiency (Mn excess)
0·005	5·00	0·0010	0·83	1·29	1·57	Very severe Fe deficiency (Mn excess)
0·500	0·000	—	5·66	15·8	1·51	Very severe Mn deficiency (Fe excess)
0·50	0·002	250·0	5·44	14·5	2·22	Severe Mn deficiency (Fe excess)
0·50	0·010	50·0	4·47	12·4	3·01	Medium Mn deficiency (Fe excess)
0·50	0·250	2·0	1·95	2·24	3·78	Normal
0·50	2·00	0·25	0·72	1·26	3·16	Slight Fe deficiency (Mn excess)
0·50	5·00	0·10	0·58	1·17	3·02	Medium Fe deficiency (Mn excess)
3·00	0·000	—	9·82	17·4	1·37	Very severe Mn deficiency (Fe excess)
3·00	0·002	1500	9·85	15·1	1·39	Very severe Mn deficiency (Fe excess)
3·00	0·010	300	9·88	13·3	1·90	Medium Mn deficiency (Fe excess)
3·00	0·250	12·0	2·48	4·15	3·78	Normal
3·00	2·00	1·5	1·50	2·51	4·05	Normal
3·00	5·00	0·6	0·76	1·77	3·70	Slight Fe deficiency (Mn excess)

manganese respectively were all normal and possessed about the same dry weight, provided that in each case the ratio of soluble iron to soluble manganese in the leaves was within the range 1·5–2·5. Whenever the ratio was outside this range pathological symptoms tended to develop.

If the ratio were above 2·5 the symptoms were of one kind, if the ratio were below 1·5 the symptoms were of a different kind. The former were thus those of a too high Fe/Mn ratio, the latter of a too low Fe/Mn ratio. The first could be described either as manganese deficiency or iron excess, the second either as iron deficiency or manganese excess. The first sign of a too high Fe/Mn ratio was a fading of the green colour of the lower leaves which later developed into an intervenal yellowing on the basal part of the leaves. Next the upper leaves showed a fading of the green colour in the intervenal areas followed by the development of small brown necrotic spots. Finally, the new leaves opened with the necrotic areas already present, and these leaves might fail to develop and fall; also stem apices might die. Roots showed no visible symptoms apart from being smaller than those of normal plants.

The symptoms of a too low Fe/Mn ratio (iron deficiency, manganese excess) were quite distinct from those just described. The first sign was a slight brown discoloration of the roots followed by yellowing and slight curling of the upper leaves. As the condition developed the discoloration of the roots and chlorosis of the upper leaves continued until the newer leaves were almost white. The leaves curled downwards and sometimes the midribs darkened and their tissue broke down. Large necrotic areas developed in the most chlorotic leaves, this being accompanied by death of the stem apices.

In both it was possible to correct the pathological symptoms by altering either the iron or manganese concentration in such a way as to bring the Fe/Mn ratio to a value between 1·5 and 2·5. A further indication of the importance of the Fe/Mn ratio was obtained by Somers, Gilbert and Shive (1942) in the respiration rate of soya beans in water cultures supplied with different proportions of the two nutrients. Respiration rates were always definitely lower when the Fe/Mn ratio was outside the range 1·5–2·5 than when it was within it.

Twyman (1951) examined the effect of varying both manganese and iron supply on the absorption of manganese by tomatoes, lettuce and oats and found that in general with the same level of manganese supply increase in the concentration of iron supplied as ferrous sulphate with tartaric acid, as ferric citrate or ferric ammonium citrate, led to a reduction in the content of manganese in the tissues. Some of Twyman's results summarized in Table 15 make this clear. Similarly the manganese content of the tissues of lettuce plants growing in a medium containing 0·250 p.p.m. of manganese showed a progressive reduction with increase in the iron content of the medium, the respective manganese contents in p.p.m. with iron supplied in concentrations of 2, 10, 30 and 50 p.p.m. being 150, 97, 57 and 34.

TABLE 15. *Effect of varying iron supply on the absorption of manganese.* (Data from Twyman)

Species	Manganese concentration in medium	Manganese concentration in tissues with different concentrations of iron in medium				
		0·005	0·010	0·050	0·500	3·0
Tomato	0·000	—	—	0·0	0·2	1·0
	0·005	—	—	2·0	3·0	2·0
	0·050	—	—	31·0	20·0	12·0
Oats	0·000	5	6	—	4	6
	0·002	12	10	—	3	2
	0·005	34	30	—	7	8
	0·010	53	54	—	9	6
	0·050	220	183	—	34	21
	0·250	529	454	—	92	66

The conclusions of Somers and Shive have received no more than partial support from subsequent investigators. Several workers including Morris and Pierre (1947) and Berger and Gerloff (1947) have regarded the symptoms of manganese toxicity as distinct from those of iron deficiency although Twyman decided that mild forms of manganese toxicity were indistinguishable from those of iron deficiency. Wallace and Hewitt (1946) recorded the occurrence of both iron deficiency and manganese deficiency in a number of plants, raspberries, apple, plum and peach, while they considered that although excess of manganese might produce chlorosis it did not do so in all plants. In some plants other symptoms might accompany the chlorosis produced by excess of manganese and the pattern of the chlorosis might be different from that produced by shortage of iron. That this was so in beans was reported by Wallace, Hewitt and Nicholas (1945). Both Nicholas (1949) and Twyman concluded that the ratio of total iron to total manganese in the plant was not very significant in determining manganese deficiency or toxicity although Twyman thought that the ratio might be important in the metabolism of healthy plants and so be a factor in determining growth and yield. A differentiation of the iron and manganese in the plant into active and inactive fractions might reveal more precise relationships. A high proportion of inactive manganese in the plant would explain why a characteristic manganese deficiency has been observed in plants containing much manganese while a low manganese content in which all the manganese was in an active condition could explain why deficiency symptoms did not develop in plants with a low total content of this element.

In a study of the effects of various nutrients on the absorption of manganese by potato plants grown in sand cultures, Bolle-Jones (1955) found that with increase in the supply of iron to the medium, which contained adequate manganese and other nutrients, the concentration of manganese increased in the roots but decreased in stems and leaves. This finding suggested that iron had the effect of reducing the mobility of manganese in the plant such that its rate of translocation from root to shoot was lowered.

The difficulty of generalizing on manganese–iron relationships is shown by comparing Twyman's results with oats, tomatoes and lettuce cited above with those of Weinstein and Robbins (1955) on sunflowers. Twyman found that increasing the iron concentration in the medium, with constant manganese supply, could induce symptoms of manganese deficiency; while, with low concentrations of iron (0·005 to 0·01 p.p.m.), increase in the concentration of manganese in the medium could lead to an increase of iron in the tissues. Weinstein and Robbins, on the contrary, found that a high concentration of iron did not induce symptoms of manganese deficiency, while manganese in high concentration did induce the chlorosis indicative of deficiency of iron.

Other cations as well as iron have been found to influence the absorption of manganese.

In a study of cation absorption by tobacco, Swanback (1939) made some observations on the effect of potassium and calcium on the absorption of manganese. The plants were grown in nutrient solutions containing calcium in three different concentrations. In the absence of manganese the symptoms of deficiency of this element were most pronounced in the plants grown in the solutions containing the highest concentration of calcium (0·403 g per litre), were less with a medium calcium concentration (0·143 g per litre), and were not observed with low calcium content (0·042 g per litre). The effects with potassium were in the reverse order, the symptoms of manganese deficiency being most pronounced with low potassium supply (0·026 g per litre), less with medium potassium supply (0·082 g per litre), and not noticeable in the plants supplied with the highest concentration of potassium (0·26 g per litre). These results are interpreted as suggesting that there is an antagonism between calcium and manganese in their absorption, while there is none between potassium and manganese. Other observations by Swanback support this suggestion. Thus with a low supply of calcium the dry matter produced by the tobacco plant cultures was six times as much when manganese was not supplied as when 0·0054 millimol per litre of this element was supplied to the culture solution. With a high supply of calcium the reverse resulted, the dry matter of the plants provided with manganese being three and a half times that of the plants not supplied with it. These results are explained on the view that with low calcium supply the manganese retards the absorption and utilization of calcium, shortage of which results in the small amount of dry matter produced, while with high calcium supply the antagonistic effect of the manganese is insufficient to reduce the absorption and utilization of the calcium to a low level while the favourable effect of the manganese itself brings about an increase in growth and so of dry matter produced.

That calcium antagonizes the absorption and utilization of manganese while potassium is indifferent is shown by determinations of the manganese content of shoots and roots of tobacco plants grown in the solutions containing the various concentrations of calcium and potassium already mentioned and 0·0054 millimol of manganese per litre. The results are shown in Table 16.

TABLE 16. *Effect of varying concentrations of calcium and potassium on the absorption of manganese by tobacco plants in water culture.* (From Swanback)

Treatment	Plant	Manganese content in millimol × 10⁻³ per g after		Manganese translocation quotient after	
		45 days	60 days	45 days	60 days
Low calcium	Shoot	0·74	0·75	0·91	1·25
	Root	0·82	0·60		
Medium calcium	Shoot	0·25	0·36	0·70	0·07
	Root	0·36	5·00		
High calcium	Shoot	0·16	0·33	0·61	0·06
	Root	0·26	5·60		
Low potassium	Shoot	0·22	0·20	0·43	0·10
	Root	0·51	2·10		
Medium potassium	Shoot	0·25	0·36	0·70	0·07
	Root	0·36	5·00		
High potassium	Shoot	0·30	0·28	0·61	0·09
	Root	0·49	3·20		

The values in the last two columns of the table are of what Swanback calls the 'translocation quotient'. This is the ratio of the manganese content of the shoot to that of the root and is supposed to give a measure of the mobility or relative translocation of the manganese. They show clearly the effect of calcium in reducing both the intake of manganese and its translocation to the leaves, while no such effect is suggested as regards potassium.

With soya beans Evans, Lathwell and Mederski (1950) found a similar effect of potassium for from culture solutions lacking this element the uptake of manganese was only 82 per cent of that from a nutrient solution containing an adequate supply of potassium.

In pineapple Sideris and Young found (1949) that ammonium salts, unlike those of potassium noted above, brought about a reduction in the uptake of manganese, apparently attributable to antagonism between the two elements in their absorption by the plants.

Bolle-Jones (1955) examined the effect of several mineral nutrients on the uptake of manganese by potato plants grown in sand cultures. His experiments with iron have already been mentioned. Increasing the potassium concentration of the medium brought about some decrease in the manganese content of the lamina of the leaves during the early stages of growth but the decrease gave place to an increase towards the end of the season. With increasing potassium concentration there was an increase in the manganese content of stems and petioles, but altogether the concentration of potassium in the medium had little effect on the manganese content of the whole plant. With increased supply of calcium as calcium carbonate the manganese content of both leaf laminae and tubers was reduced, while with increase in the phosphate content of the medium the uptake of manganese was lessened and the content of manganese generally throughout the plant was less, stems, roots, petioles and leaf laminae all being affected in this way.

ZINC

Little, if anything, appears to be known of the concentration of zinc in the soil solution, but it is likely to be extremely low as the zinc compounds occurring naturally in soils are insoluble in water. Determinations of the exchangeable zinc in soils indicate values only up to about 23 p.p.m., much less than the highest values found for manganese. The zinc extractable with 0.1N hydrochloric acid in fifty-eight soils of West India was found by Nair and Mehta (1959) to range from 0.50 to 6.05 p.p.m., with an average value of 3.06 p.p.m.

The analyses of the soils of Finland made by Lounamaa showed that generally these contain much less zinc than manganese, most of the values lying between 100 and 600 p.p.m. although in a few soils a content of 1000 p.p.m. was found. In the soils of West India mentioned above Nair and Mehta found that the total zinc content ranged from 20 to 95 p.p.m. while in the West Pakistan soils examined by Wahhab and Bhatti (1958) the total zinc content mostly ranged from 15.0 to 87.5 p.p.m.; a soil with only a trace of zinc was exceptional, but low values have also been recorded elsewhere, as for instance with Scottish peats in which values as low as 10 p.p.m. were found by Mitchell. Values not much higher than this, namely, 14 and 16 p.p.m. of the oven-dry soil, were found by Connor, Schimp and Tedrow (1957) for two podsolic soils in the United States. The values for calcareous soils, 169 and 320 p.p.m. and for two soils formed on glacial drift, 102 and 86 p.p.m., were all within the range recorded by Lounamaa.

The zinc content of plants is on the average lower than that of manganese. Thus the zinc content of the ash of the plants, apart from lichens, analysed by Lounamaa ranged for the most part from 100 to 3000 p.p.m., though a few contained less than 100 p.p.m. and a few as much as 6000 p.p.m., while in one instance a moss *Rhacomitrium lanuginosum* was recorded as containing 10 000 p.p.m. of zinc in the ash, although samples of the same moss from other localities contained only 1000 and 3000 p.p.m. respectively. Lichens appear to be characterized by a high zinc content, the range of values found for them by Lounamaa being 1000 to 10 000 p.p.m. of ash, and in lichens on the average the content of zinc was found to be higher than that of any other trace element as the data given in Table 17 indicate.

TABLE 17. *Average trace-element content of lichens in p.p.m. of ash.* (Data from Lounamaa)

Element	On siliceous rocks	On ultrabasic rocks
Manganese	410	1800
Zinc	3800	2500
Boron	130	260
Copper	280	280
Molybdenum	19	38

This was also the case with mosses growing on calcareous soils, while with those growing on soils overlying siliceous and ultra-basic rocks the contents of manganese and zinc were of about the same order. This is in marked contrast with the state of affairs in vascular plants, in which the content of zinc is generally much less than that of manganese and of which the values for leaves of herbaceous flowering plants may be taken as representative (see Table 18).

As with manganese the amount of zinc absorbed by plants of the same species may show a wide variation. For example, Lounamaa found the zinc content of the ash of different plants of ling (*Calluna vulgaris*), all from siliceous habitats, ranged from 100 to 3000 p.p.m. As Table 18 shows, on the whole more zinc is taken up by plants on siliceous soils than from either ultrabasic or calcareous soils. As will be noted below this is probably related to the reaction of the soil.

TABLE 18. *Average trace-element content of the leaves of herbaceous flowering plants in p.p.m. of ash.* (Data from Lounamaa)

Element	On siliceous soils	On ultrabasic soils	On calcareous soils
Manganese	7200	3700	2800
Zinc	930	330	270
Boron	640	980	590
Copper	130	93	73
Molybdenum	13	22	17

That the content of zinc in plants is influenced by its concentration in the medium was demonstrated by the experiments of Rogers and Wu (1948) with oats grown in sand cultures supplied with different amounts of zinc sulphate. From the data presented in Table 19 it will be seen that as the content of zinc in the medium was increased so was the zinc content of the plant.

TABLE 19. *Effect of concentration of zinc in the medium on the zinc content of oat plants.* (Data from Rogers and Wu)

Zinc in the medium (lb of $ZnSO_4.7H_2O$ per acre)	Zinc content of plants in p.p.m. of dry weight	
	Young plants	Older plants
0	7·8	6·1
50	12·8	11·2
100	15·2	11·6
200	30·3	14·3

The amount of zinc available for plants is dependent on the acidity of the medium, decreasing with increase in pH. As a result of this, zinc deficiency is likely to occur on alkaline or even neutral soils and indeed Camp (1945) concluded that with *Citrus* deficiency of zinc could be expected if the pH of the soil exceeded 6. However, in the experiment of Rogers and Wu, the results of which are summarized in Table 19, the pH of the medium ranged from 6·46 to 6·80, just a very little on the acid side of neutrality. When, however, the pH was increased by dressings of calcium carbonate Rogers and Wu found the zinc content of oat plants decreased, with one exception among their determinations, as the amount of added calcium carbonate and the pH increased. In the experiment the results of which are summarized in

Table 20 the medium contained 50 lb of zinc sulphate ($ZnSO_4$. $7H_2O$) per acre and calcium carbonate was applied in amounts ranging from 100 to 2000 lb per acre. The effect of this treatment in reducing absorption of zinc is clearly seen from the analyses of the young plants but is less pronounced although still obvious, apart from one value, in the older plants.

TABLE 20. *Effect of addition of calcium carbonate to the medium on the uptake of zinc by oat plants.* (Data from Rogers and Wu)

Calcium carbonate added (lb per acre)	Zinc content of plants (p.p.m. of dry weight)		pH of medium	
	Young plants	Older plants	Young plants	Older plants
100	19·5	17·4	5·88	6·10
500	7·7	12·2	6·92	6·90
1000	5·7	8·5	7·28	7·39
2000	4·7	10·5	7·60	7·52

Rogers and Wu also found that the uptake of zinc was reduced by addition of calcium hydrogen phosphate to the medium. With 50 lb of zinc sulphate per acre added to the medium dressings of phosphate ranging from 100 lb (as P_2O_5) to 1500 lb per acre were applied. In the younger plants increasing the dressing from 100 to 500 lb per acre brought about a decrease in the zinc content of the young oat plants from 30·1 p.p.m. to 17·0 p.p.m. and of the older plants from 21·6 to 17·6 p.p.m. With further increase in the phosphate content of the medium there was little further change in the zinc content of the plants. Here the pH was not significantly affected by increased addition of the phosphate, the measured values ranging from 5·45 to 5·68 in the culture medium of the younger plants and from 5·63 to 5·91 in that of the older plants. These results might be related to phosphate in some way not understood, or might result from an antagonism between the zinc and calcium ions. From the work of Wear (1956), however, the latter explanation would not appear to be likely. Wear grew plants of *Sorghum* in soil cultures in a greenhouse, the soil used being a sandy loam with a pH of 5·6. To the various cultures calcium carbonate, sodium carbonate or calcium sulphate was added in a range of levels combined with zinc sulphate in various amounts. The remaining major and minor nutrients were provided in appropriate quantity. The levels of calcium carbonate were 0, 2000, 4000 and 8000 lb per acre, of sodium carbonate 0, 500, 1000 and 2000 lb per acre, and of calcium sulphate the same. Zinc sulphate was provided at the levels of 0, 20 and 40 lb per acre.

Application of calcium carbonate at the rate of 2000 lb per acre increased the pH of the soil to 6·6 and the uptake of zinc at all levels was greatly reduced. The calcium content of the plants, however, increased from 0·78 to 1·09 per cent. Greater dressings of calcium carbonate brought about a further, though slighter, fall in zinc uptake with the further increase in soil pH. Sodium carbonate also brought about a rise in soil pH which was accompanied by a fall in zinc intake. Application of calcium sulphate at the rate of 2000 lb per acre brought about a decrease in soil pH to 4·8 while the zinc content of the plant was increased slightly. The calcium content of the plant was increased from 0·78 to 1·01 per cent. The uptake of zinc thus appeared to be clearly related to the pH of the soil and not to the level of calcium as such.

The effect of calcium hydrogen phosphate on the uptake of zinc by young plants of sour orange recorded by Bingham, Martin and Chastian (1948) was very similar to its effect on oats as described by Rogers and Wu. Thus the zinc content of leaves of the young plants growing in a loamy fine sand without added phosphate was found to be 28 p.p.m. When phosphorus as calcium hydrogen phosphate at the rates of 76, 360 and 900 lb per acre was added to the soil the zinc contents of the leaves were respectively 25, 18 and 20 p.p.m. With a sandy loam the corresponding zinc contents of the leaves of the young *Citrus* plants were 35 p.p.m. without added phosphate and 33, 22, 27 and 27 with additions of calcium hydrogen phosphate to the soil in amounts equivalent to 76, 360, 900 and 1800 lb of phosphorus per acre respectively. The tentative suggestion was made that this effect of phosphate was due to the production of relatively insoluble zinc phosphate in the soil. It was considered not to be a calcium effect as zinc-deficiency symptoms were not produced on soils with high concentrations of calcium in some other form.

BORON

Natural soils generally contain very much less boron than manganese or even zinc. Lounamaa found the average boron content of the siliceous, ultrabasic and calcareous soils of Finland examined by him to be 58, 41 and 78 p.p.m. respectively. Some soils were found to contain less than 3 p.p.m. while a few contained more than 100 p.p.m. but only one of the soils analysed contained as much as 300 p.p.m. Comparable ranges of boron content were found in soils of India (7 to 100 p.p.m.) by Ramamoorthy and Viswanath (1946) and in soils of Czechoslovakia (8 to 75 p.p.m.) by Koppova and Duchon as cited by Lounamaa. For some cultivated soils examined by Philipson (1953) the boron content ranged from 21·4 p.p.m. for a loam to 38 p.p.m. for a clay.

Since the actual content of boron in soils is generally low and the soil minerals containing boron practically insoluble in water, the concentration of boron in the soil solution is likely to be very low indeed. Attempts at estimating the available boron have been made by arbitrary methods such as extracting the soil with hot water or phosphoric acid, the latter extractive generally giving higher values than the former. For cultivated soils in Sweden Philipson found the content of boron extracted with phosphoric acid ranged from 0·76 p.p.m. for a clay to 4·04 p.p.m. for a loamy sand.

Although the boron content of most soils is so much less than that of zinc, plants may absorb as much boron as zinc or even more (cf. Table 18). According to Lounamaa's analyses woody plants appear to acquire a higher content of boron than herbaceous plants, while lichens and grasses absorb on the whole less boron than other plants. As with other trace elements the amounts absorbed by different plants of the same species may differ considerably, even when they are growing near one another in the same habitat. Thus Lounamaa called attention to the boron content of the leaves and twigs of *Betula verrucosa* in plants of which growing in the same habitat the boron content of the ash ranged from 300 to 3000 p.p.m. Actually as wide a divergence in the manganese content was shown by the leaves of these same plants.

As is to be expected, increase in the boron concentration of the medium brings about an increase in the boron content of the plant.

Some numbers illustrating this, taken from the results of an investigation of the calcium–boron interrelationships of Siberian millet (*Setaria italica*) by McIlrath and de Bruyn (1956), are given in Table 21.

TABLE 21. *Effect of the level of boron supply on the absorption of boron by* Setaria italica. (Data from McIlrath and de Bruyn)

Boron supply (p.p.m.)	Boron content (mg per g dry weight of shoots) with calcium supplied at the rate of	
	40 p.p.m.	480 p.p.m.
0·0	0·029	0·014
0·05	0·020	0·018
0·5	0·043	0·034
5·0	0·262	0·194
50	1·166	0·760

Bingham, Martin and Chastain (1958) found that in eight out of nine Californian soils addition of calcium hydrogen phosphate resulted in a reduced uptake of boron by young sour orange plants. Their results with two of these soils, a loamy fine sand and a sandy loam, are summarized in Table 22.

TABLE 22. *Effect of calcium hydrogen phosphate on the uptake of boron by* Citrus (*sour orange*). (Data from Bingham, Martin and Chastain)

Calcium hydrogen phosphate added as lb P per acre	Boron content of leaves (p.p.m.)	
	loamy fine sand	sandy loam
0	34	67
76	31	59
360	25	61
900	22	41
1800	—	30

It is not clear whether this result is to be ascribed to the effect of calcium as found by McIlrath and de Bruyn (p. 136) or to the effect of phosphate such as Bingham and his co-workers found with zinc and copper.

It is generally held that with increase in the pH the solubility and availability of boron in the soil is reduced and so liming the soil, by increasing the pH, reduces the absorption of boron by plants. This view has, however, not been universally accepted. Thus Drake, Sieling and Scarseth (1941) concluded that the solubility of boron was unaffected over the pH range 4·1 to 11·6. It may thus be that effects resulting from liming may not be due to an increase in pH but to an effect of calcium ions on the absorption of boron.

Indeed, several investigations have indicated that both potassium and calcium influence the uptake of boron. In this connexion the work of Reeve and Shive (1944), who examined the relations of potassium and calcium to boron in tomato, may be particularly noted. In their experiments on the relation of potassium to boron twenty cultures, each containing three plants, were grown in sand cultures which received a nutrient solution supplied in a continuous flow. Five different potassium concentrations were used, namely, 10, 50, 89, 250 and 500 p.p.m., there being thus four cultures receiving potassium in each one of these concentrations. The four cultures at each potassium level respectively received boron in the concentrations 0·001, 0·1, 0·5 and 5·0 p.p.m. The nutrient solution contained iron, manganese and zinc in addition to the major nutrients.

It was found that symptoms of boron deficiency appeared first in the culture supplied with 0·001 p.p.m. of boron and 500 p.p.m. of potassium, and last in the culture supplied with 0·001 p.p.m. of boron and 10 p.p.m. of potassium and that, in general, the severity of the symptoms of boron deficiency increased with increase in the potassium concentration. No symptoms, either of boron deficiency or excess, were observed in any of the cultures receiving the intermediate concentrations of boron (0·1 and 0·5 p.p.m.), but all cultures receiving 5 p.p.m. of boron developed symptoms of boron toxicity. With these the severity of the symptoms increased with increase in potassium concentration. Thus increasing potassium concentration brought about a progressive increase in the severity of the symptoms of boron deficiency in low boron concentrations and of the symptoms of boron excess in high boron concentrations.

Analysis of the plants showed that at each level of supplied boron the amount of both soluble and total boron in the tissues increased with increase in the potassium concentration. This explained the accentuation at the high boron level of the symptoms of boron toxicity with increasing potassium concentration, but did not explain why the boron deficiency symptoms at the low boron level should be accentuated with increase in the potassium supply.

In their experiments on the relation of calcium to boron thirty cultures were used in which six levels of calcium supply (5, 10, 50, 100, 250 and 500 p.p.m.) and five of boron (0·001, 0·01, 0·5, 5·0 and 10·0 p.p.m.) were employed. The results showed that calcium acted similarly to potassium in that with the lowest boron concentration (0·001 p.p.m.) the severity of the symptoms of boron deficiency increased with increase of the supply of calcium, the effect of the latter in this respect being, indeed, greater than that of potassium. In its influence on boron toxicity produced by the highest boron concentration (10 p.p.m.), however, calcium acted in exactly the opposite way to potassium, for with progressively increasing calcium supply the severity of the symptoms of boron toxicity became less. Chemical analyses of the plants showed that with the lower concentrations of boron in the nutrient solution (0·001, 0·01 and 0·5 p.p.m.) the boron content (both total and soluble) was independent of the calcium concentration of the nutrient solution, but with the higher concentrations of boron in the nutrient solution (5·0 and 10·0 p.p.m.) increase in the concentration of calcium brought about a decrease of both total and soluble boron in the plants. This accounted for the effect of increasing concentration of calcium in reducing the toxicity of boron when supplied in high concentration. The ratio of calcium to boron in the plant was influenced by the supply of potassium, increase of this cation in the nutrient solution bringing about a lowering of the calcium/boron ratio. Calcium appeared to have no significant influence on the potassium/boron ratio.

The results obtained by Reeve and Shive are in harmony with the well-known fact that heavy liming of certain soils will induce boron deficiency in a number of crop plants such as beet and swede. It will also be observed that if a soil contains sufficient boron to induce toxicity symptoms, heavy liming should reduce the severity of these. This was found to be so with oats by Jones and Scarseth (1944). These workers grew lucerne, oats and tobacco in pots of limed and unlimed soils to which various quantities of borax were added. The calcium and boron in the plants were determined. As a result the conclusion was drawn that a plant will only make normal growth when there is a certain balance between the intake of calcium and that of boron. From a consideration of their own data and those obtained by Cook and Miller (1939) for sugar beet, Muhr (1940) for soya bean and Drake, Sieling and Scarseth (1941) for tobacco, they came to the conclusion that the ideal balance for these various plants is attained when the ratios (in equivalents) of calcium to boron in the respective plants are 100 for sugar beet, 500 for soya bean and 1200 for tobacco.

A study of the interrelationships of boron and calcium in Siberian millet (*Setaria italica*) was made by McIlrath and de Bruyn (1956). Plants grown in sand culture were supplied with boron as boric acid in concentrations of 0, 0·05, 0·5, 5 and 50 p.p.m. and with calcium as calcium nitrate or, in the higher levels, as mixtures of calcium nitrate and calcium chloride, in concentrations of 40, 160, 320 and 480 p.p.m. The effect of increasing calcium supply was to bring about a decrease in both the total and soluble boron in the shoots. This was not very marked with boron supplied at lower levels but was more definite at higher levels of boron supply, as the results summarized in Table 23 indicate.

TABLE 23. *Effect of the level of calcium supply on the absorption of boron by Siberian millet.* (Data from McIlrath and de Bruyn)

Calcium supply (p.p.m.)	Total boron as mg per g dry weight of shoots in plants supplied with		Soluble boron as mg per g dry weight of shoots in plants supplied with	
	0·05 p.p.m. B	5 p.p.m. B	0·05 p.p.m. B	5 p.p.m. B
40	0·020	0·262	0·009	0·231
160	0·026	0·278	0·011	0·249
320	0·016	0·231	—	0·211
480	0·018	0·194	0·012	0·188

The length of the daily period of illumination of soya bean plants was found by MacVicar and Struckmeyer (1946) to be a factor influencing the effect of boron deficiency. Seedlings in pot cultures both with and without a supply of boron were subjected to short-day (9 h) or long-day (17–18 h) illumination. Three weeks after the plants were first subjected to these photoperiods no visible differences were detected in the short-day plants between those with and without a boron supply. It was different with the long-day plants. With these, reduced growth of the boron-deficient plants was already obvious, and a month after the transference to the long-day condition the plants exhibited typical symptoms of boron deficiency. After a further month these symptoms were very severe, whereas in the boron-deficient short-day plants the symptoms were only moderately severe.

COPPER

The mean copper contents of the siliceous, ultrabasic and calcareous soils of Finland examined by Lounamaa were found to be respectively 110, 22 and 42 p.p.m. Siliceous soils thus contained on the average about twice as much copper as boron, while the other soils contained roughly only half as much. Average values obtained by Stenberg, Ekman, Lundblad and Svanberg for mineral and organic soils in Sweden and cited by Wiklander (1958) were respectively 16 and 19 p.p.m., while in West Pakistan Wahhab and Bhatti (1958) obtained values up to 77 p.p.m. with an average of 45 p.p.m. Low values, down to 3·3 p.p.m., cited by Wiklander, were found by Hoffmann for peat soils in Germany. In two podsolic soils in the United States Connor, Schimp and Tedrow (1957) found only 4 p.p.m. of copper in the oven-dry soil, while in two soils derived from calcareous material the amounts were 13 and 17 p.p.m., and in two soils formed on glacial drift the amounts were 40 and 30 p.p.m. Mitchell (1955) gave the range 3 to 100 p.p.m. as covering the copper content of most mineral soils.

According to Wiklander only about one per cent of the copper in the soil is soluble in water and the concentration of copper in the soil solution of ordinary soils is only about 0·01 p.p.m. Little appears to be known about the availability of copper in soils, but by the use of the bioassay technique Nicholas and Fielding (1947) estimated the content of available copper in an alkaline fen soil from Methwold, Suffolk, as 0·5 to 0·75 p.p.m., while in a calcareous drift soil from Long Ashton it was more than 5 p.p.m.

The exchangeable copper is strongly adsorbed by the soil colloids and particularly by organic matter with which, indeed, it is also supposed to form stable complex compounds, in which form it is non-exchangeable. Steenbjerg (1950) considered that the combination of cupric ions with humus was one of the most important factors in rendering copper unavailable to plants and so inducing copper deficiency.

The amount of available copper appears to be affected by the pH of the soil. According to Steenbjerg, and also Gaarder and Røyset, quoted by Wiklander, the copper content of oats was found to be at a minimum with a pH of the soil of 5·5.

While the range of copper content of plants is wide it is, on the whole, less than with other trace elements. The copper content of the ash of the leaves of fifteen species of angiosperms according to Erkama (1950) ranged from 60 to 240 p.p.m. These values agree well with those obtained by Lounamaa, of which 93 per cent of more than 400 analyses relating to 83 species of herbaceous angiosperms lay between 30 and 300 p.p.m. of the ash. Beeson, Gray and Adams (1947) found the copper content of a number of grasses growing on the same sandy loam ranged from 21 p.p.m. in Kentucky blue grass to 4·5 p.p.m. in para grass.

No doubt as with other trace elements the absorption of copper increases with the concentration of this element in the medium. This was indicated by analyses made by Mulder (1950) of plants grown on soils inducing copper deficiency and on the same soils to which copper sulphate had been added. For example, the copper content of the shoots of oat plants five weeks old grown on a copper-deficient soil was 1·9 p.p.m. of the dry weight and the plants showed signs of copper deficiency; the copper content of the shoots of plants grown on the same soil to which 50 kg of copper sulphate per hectare had been added was 8·1 p.p.m. of the dry weight and the plants were normal. Grain from wheat plants showing slight symptoms of copper deficiency grown on a sandy soil had a copper content of 0·9 p.p.m. of the dry weight, while in grain from normal plants on the same soil to which 25 mg of copper sulphate per kilogram of soil had been added the copper content was 3·0 p.p.m. of the dry weight.

Bingham, Martin and Chastain (1958) found the uptake of copper by sour orange was markedly depressed by the addition of calcium hydrogen phosphate to the soil. Thus the copper content of the leaves of young plants growing on a sandy loam without added phosphate was 10 p.p.m. With addition of calcium phosphate in amounts equivalent to 76, 360, 900 and 1800 lb of phosphorus per acre the copper contents were reduced respectively to 4·2, 2·3, 1·7 and 0·9 p.p.m. As with zinc, with which the depressing effect of phosphate on uptake is much less marked, the suggestion was made that insoluble copper phosphate might be formed external to the root, thus rendering the copper non-available.

The effect of the presence of other salts on the toxicity of copper to white lupins (*Lupinus albus*) was examined as long ago as 1903 by True and Gies who found that different salts had different effects. The presence of sodium chloride was found to increase slightly the toxic effect of cupric chloride while calcium chloride reduced it and magnesium chloride had no effect. According to Kahlenberg and True (1896) the toxicity of copper towards white lupins is also reduced in the presence of sugar and potassium hydroxide.

That the presence of other substances in the medium can modify the toxic effects of excess copper could be due to a reaction of the substance with copper in the medium reducing its availability, to an influence on the absorption of copper, or to some interaction with copper in the plant. These various possibilities still require examining.

MOLYBDENUM

Many determinations of the molybdenum content of soils in different parts of the world have been made in recent years. These show that generally the molybdenum content is very much less than that of other recognized essential trace elements. Analyses of soils in Great Britain, Finland, Russia and North America all indicate that the molybdenum content of soil is generally only a few parts per million. A number of soils in New Zealand were found by Kidson (1954) to contain less than 1 p.p.m. of dry soil. Exceptionally, considerably higher contents have been found. Thus while Lounamaa found nearly all the soils in Finland he examined contained less than 10 p.p.m. of molybdenum, one soil contained about 300 p.p.m. but that was near a molybdenite mine. Bertrand (1940) found the molybdenum content of soils in France ranged from 4·3 to 69 p.p.m. The soils of Somerset bearing the so-called 'teart' pastures were found by Lewis (1943a) to have an unusually high content of molybdenum, namely, 20 to 100 p.p.m. The soils of Hawaii appear to have an unusually high molybdenum content, Fujimoto and Sherman (1951) finding these ranged from 8·9 to 73·8 p.p.m., the average of 79 samples being 25·8 p.p.m.

With the content of molybdenum in the soil being generally so low it is to be expected that its concentration in the soil solution will be very low indeed. Barshad (1951a) reported that the water-soluble molybdenum of soils in California ranged from 0·3 to 3·9 p.p.m. of dry soil, but these contents seem to be unusually high. Values for exchangeable molybdenum in Hawaiian soils, determined by Fujimoto and Sherman (1951) by extracting with ammonium acetate, ranged from zero to only 0·13 p.p.m.

Using the bioassay method, Nicholas and Fielding (1950) found the available molybdenum in a soil at Long Ashton derived from New Red Sandstone was more than 0·5 p.p.m. of air dry soil while in a soil derived from the Devonian at St Ives, Cornwall, it was only from 0·015 to 0·03 p.p.m., and in a Lower Greensand soil from Bromham, Wiltshire, it was only 0·01 p.p.m. On these latter soils cauliflowers developed the whiptail symptoms of molybdenum deficiency, but on the first-named soil the plants were free from these symptoms.

Molybdenum, alone of the trace elements, becomes more available in soil with increasing pH, for as Barshad (1951a) has indicated, the water soluble molybdenum of the soil increases with increasing pH. According to Amin and Joham (1958), as well as soluble molybdates, molybdenum occurs in soils as three oxides, molybdenum trioxide MoO_3, molybdenum dioxide MoO_2 and a third oxide Mo_2O_5. The soluble molybdates are readily available to plants, and the trioxide, while not available as such, is readily rendered so by reaction with cations in the soil. The two lower oxides are not available. Hence oxidation of the lower oxides tends to increase the availability of molybdenum while reduction tends to diminish the availability. The effects of oxidation and reduction on availability are thus the reverse of the effects of such reactions in regard to the availability of manganese. Immediate availability depends on the production of soluble molybdates from molybdenum trioxide which, as we have seen, is favoured by increasing pH. Acidifying the soil will, of course, have the reverse effect, so that by so doing the toxic effects of molybdenum in plants growing on soils where it is present in excessive quantity can be alleviated. Further reference will be made to this possibility in a later chapter.

The contents of molybdenum in plants are for the most part low. For a number of New Jersey soils with molybdenum contents ranging from 0·8 to 3·3 p.p.m. the molybdenum content of lucerne (alfalfa) was recorded by Evans and Purvis (1951) as ranging from less than 0·1 to 1·4 p.p.m. of the dry weight. Fujimoto and Sherman (1951) found the molybdenum content of the vegetation on the Hawaiian soils they examined ranged from zero to 2·5 p.p.m. of the dry weight. There was no correlation between the molybdenum content of the plants and of the soils bearing them. Lounamaa found a wide variation in the molybdenum content of plants in Finland. While in some there was no measurable amount of molybdenum by the method used, in many he found 10, 20, 30 or more p.p.m. of the ash. A few values of over 100 p.p.m. were found but these were rare. Since the ash content of the dry matter of plants varies so greatly it is not possible to make any direct comparison of Lounamaa's results with others given above on a dry-weight basis, but if the ash content is taken as 15 per cent of the dry weight most of the molybdenum contents he found were of the same order of magnitude as those found by other investigators.

It is quite clear that different species have different capacities for absorbing molybdenum. Lewis (1943b) found the clovers (*Trifolium* spp.) and the grass Yorkshire fog (*Holcus lanatus*) growing on a molybdenum-rich soil contained much more molybdenum than other grasses growing in the same pastures (see Table 34, p. 187). Barshad (1948) also concluded that members of the Leguminosae had a greater capacity for accumulating molybdenum than plants of some other families.

As with other nutrients absorption of molybdenum increases with increase in the level of its supply. An experiment of Fujimoto and Sherman may be quoted to illustrate this. Sudan grass grown in pots of soil was treated with ammonium molybdate in various quantities ranging from 2 to 2000 lb per acre. The molybdenum contents of the crops on three soils described as reddish brown, dark magnesium clay and low humic latosol are shown in Table 24.

TABLE 24. *Effect of supplying molybdate at different levels to the molybdenum content of Sudan grass.* (Data from Fujimoto and Sherman)

Ammonium molybdate supplied (lb per acre)	Molybdenum content of crop (p.p.m.) on three soils		
	Reddish brown	Dark magnesium clay	Low humic latosol
0	3·4	4·0	3·2
2	10·0	4·0	3·6
20	62·0	28·0	3·4
200	298·0	225·0	15·4
2000	—	525·0	30·0

No doubt the absorption of molybdenum is affected by the presence of other substances in the medium. As regards calcium carbonate it has already been pointed out that by increasing the pH the molybdenum in the soil is rendered more available. That this does indeed result in an increase in the uptake of molybdenum is shown by the observation of Barshad (1951b) who found that by the addition of calcium carbonate to an acid soil with pH 4·7 the molybdenum content of plants on it was raised from 7·4 to 23·2 p.p.m.

Deficiency of sulphur in the medium of seedlings of rubber (*Hevea brasiliensis*) grown in sand culture was found by Bolle-Jones (1956) to lead to increased molybdenum content of the leaves. The molybdenum contents of sulphur-deficient leaves was 3·38 p.p.m. of the dry weight, while that of the leaves of plants adequately provided with sulphur was only 1·40 p.p.m.

CHAPTER VI

THE FUNCTIONS OF TRACE ELEMENTS
IN PLANTS

THAT the trace elements are required by plants in very small amounts suggests that their function may be that of catalysts in vital reactions. At the same time it does not follow that this is their only function; they may form part of substances essential to the life of the plant other than catalysts but which are necessary in only small quantity. It is, however, now clear that all the generally recognized trace elements with the possible exception of boron, either form an integral part of certain enzymes or activate enzyme systems, at any rate when these are isolated from the living plant. Under such conditions it is frequently found that more than one metal may be capable of activating the same enzyme action and when this is so it would appear that no one of these can be regarded as essential for that particular action. Also it has to be borne in mind that the functioning of an element in the plant where many reactions are linked and interlocked may not be precisely indicated by the part it plays in an isolated action outside the plant. Nevertheless examination of the action of various elements, in activating enzyme actions in plant extracts and other dead plant material, has given valuable information on the functions of trace elements and in some instances is the most definite we have.

As well as the part trace elements play in enzyme actions a variety of other functions have been ascribed to them. This is particularly so with manganese, zinc and boron but less so with copper and molybdenum. In this connexion it may possibly be significant that the first three trace elements are generally present in plants in a significantly higher concentration than the other two (cf. Table 18, p. 129).

A point worthy of note is that the effects of boron deficiency are generally of quite a different kind from those of the other trace elements. Warington (1923) thought that the function of boron in the broad bean was nutritive rather than catalytic.

With boron deficiency, as we have seen, the most characteristic effect is a breakdown of thin-walled tissues, especially those of the meristematic regions, followed by degeneration of the vascular tissues, whereas with the other four well-established trace elements early external symptoms are generally localized chloroses, although these may precede more serious disturbances in growth such as those which lead, in the case of zinc and copper, to die-back of the terminal buds of the shoots. Nevertheless, chlorosis may also be a symptom of boron deficiency, as, for example, in tobacco (Van Schreven, 1934) and in sugar cane (Martin, 1934), while plants do not appear to require more boron than manganese. It is not very clear why a catalyst essential for the maintenance of normal metabolism and growth should not be regarded as fulfilling a nutritive function.

A suggestion was made some twenty-six years ago by Thatcher (1934) that the functions of various elements concerned with plant metabolism were related to their position in the periodic classification table. In the first place he pointed out that nearly all the elements which are known or have been suggested to have a function in plant nutrition are included in the first four periods or orbits of the periodic classification. These first four periods, with the atomic numbers* of the elements, are shown in Table 25. Of the elements shown in this table we may dismiss the inert gases of group O. In the other groups there are a number of elements which so far have not been found essential for plants, but all the elements known to be necessary for plants are included in these first four periods with the one exception of molybdenum (period 5, group VI, atomic number 42). The heavier metals do not figure among elements necessary for plants.

Thatcher's classification of the elements in relation to their functions in the plant was based on the conception that 'green plants are the energy-absorbing and energy-storing agents of the cycle of life'. From this point of view he divided the elements into eight groups as shown in Table 26.

It is not within the scope of the present discussion to consider the value of this scheme of classification, but we may note that

* The atomic number is the number of free positive charges in the nucleus of the atom. Except for hydrogen it is very roughly half the atomic weight.

TABLE 25. *The first four periods of the periodic classification*

Group

Period	I	II	III	IV	V	VI	VII	VIII			O
1	H 1	—	—	—	—	—	—	—			He 2
2	Li 3	Be 4	B 5	C 6	N 7	O 8	F 9	—			Ne 10
3	Na 11	Mg 12	Al 13	Si 14	P 15	S 16	Cl 17	—			A 18
4	K 19	Ca 20	Sc 21	Ti 22	V 23	Cr 24	Mn 25	Fe 26	Co 27	Ni 28	
	Cu 29	Zn 30	Ga 31	Ge 32	As 33	Se 34	Br 35	—	—	—	Kr 36

TABLE 26. *Thatcher's classification of elements in plants*

Group	Type	Function	Elements
I	—	Energy exchange	H, O
II	Anion (or acid) group formers	Energy storers	C, N, S, P
III	Cation (or base) formers with fixed valency	Translocation regulators	Na, K, Ca, Mg
IV	Cation (or base) formers with varying valency	Oxidation-reduction regulators	Mn, Fe (Co, Ni), Cu, Zn
V	Ampholytes with varying valency	Functions unknown	B, Al, Si, As, Se
VI	Anion (or acid) formers with fixed valency	Functions unknown	Cl, F (Br, I)
VII	Cation (or base) formers with varying valency	Perhaps those of group IV	Co, Ni
VIII	Ampholytes	Functions unknown	Ge, Ga and other rare elements

three of the well-established trace elements, manganese, copper and zinc, are placed, along with iron, itself often classed as a trace element, in one group, that of oxidation-reduction regulators. The fourth, boron, is placed in another group, one with unknown function, that of ampholytes of varying valency. At the time of publication of Thatcher's paper molybdenum had not been recognized as a plant trace element. Further, it is not clear why the ampholytes of Thatcher's group VIII, which included gallium, were separated from those of his group V, since in both the functions are listed as unknown. However this may be, Thatcher's classification divides the trace elements into two groups, the oxidation-reduction regulators including manganese, copper and zinc, and also iron, the other including boron of unknown function. With regard to the oxidation-reduction regulators Thatcher pointed out that if nickel and cobalt are included the group contains six elements with consecutive atomic numbers, namely, 25–30. It is interesting to note that gallium, the essentiality of which for plants has been suggested, is the next element in the series with an atomic number of 31. This means that each successive member of the series differs as regards atomic structure from the element before it only by the addition of one electron to the proton nucleus. Whether this close connexion between these elements really has any significance in respect of their physiological function it would be premature to say, but it is a fact that several of them have rather similar chemical properties. Thatcher expressed the opinion that there was sufficient evidence to justify the opinion that manganese and iron on the one hand, and copper and zinc on the other, were pairs of 'mutually co-ordinating catalysts for oxidation-reduction reactions', the former pair for reactions in which the addition or removal of oxygen is involved, the latter for reactions which concern the transference of hydrogen. As we shall see in the sequel there is a certain amount of experimental evidence in favour of at least one of these hypotheses.

Frey-Wyssling (1935) also attempted to find a relation between essentiality and position of elements in the periodic table. He used a table in which group O appeared both on the extreme left and the extreme right, but with each element of the group one period higher in the right than in the left column. If a line is then drawn through the table from argon on the left to carbon and then on to argon on the right, this line passes through or near the positions in the table of all the essential elements with the exception of hydrogen. This line Frey-Wyssling called the nutrient-line. This relationship expresses with greater precision the fact pointed out by Thatcher, that all the essential elements occur in the first four periods of the periodic table. The farther the position of an element is from the nutrient-line the more toxic it is in general. It will be observed that essential elements occur in all groups I–VIII.

A discussion on the relation between the biological essentiality of the elements and their atomic structure was contributed by Steinberg (1938c). When considered from the point of view of their position in the periodic table Steinberg concluded that three and no more than three essential elements were to be found in each group. Where less than three essential elements were known to exist in a group there was a presumption that the missing ones were yet to be found. However, Steinberg considered that correlations between essentiality and atomic structure could best be shown by tabulating the elements on the basis of their transition subshell, those in which the electron numbers have undergone a regular change in the formation of the elements and which largely determine the chemical properties of the latter. From such considerations Steinberg made certain deductions regarding the relationship between essentiality and atomic structure and of the possible essential elements not yet recognized as such. Thus the non-essentiality of silicon was indicated, while the possibility of columbium as an essential element was suggested. Steinberg's conclusions only claimed to be very tentative.

Since deficiencies of manganese, zinc and copper characteristically induce chlorosis, it is understandable that they have been held to be concerned in chlorophyll formation (cf. McHargue 1923, 1926a; Bishop, 1928). They have also been thought to act as catalytic agents in photosynthesis (see, for example, Stoklasa, 1911; McHargue, 1923; Dufrénoy and Reed, 1934). Support for this view was forthcoming from work by Emerson and Lewis (1939), who found that when the trace elements of Arnon's groups A 4 and B 7 (see p. 11) were added to the culture medium of *Chlorella pyrenoidosa* not only was the rate of growth of the alga increased but the amount of photosynthesis per unit quantity of light absorbed was also increased. The trace elements were added in two groups, those of Arnon's A 4 group + molybdenum, now denoted by the symbol A 5, and those of Arnon's B 7 group without molybdenum, this group now being denoted B 6. The addition of the A 5 group was more effective than the B 6 group, and the addition of A 5 + B 6 more effective than either alone. It would thus appear that the trace elements play some part in the photosynthetic process, probably as essential constituents of enzyme systems involved.

Some observations of Steinberg (1942) may be mentioned here. Starting from the observation that when nitrogen was supplied only as nitrate the growth of the fungus *Aspergillus niger* was lessened in air deprived of carbon dioxide, he examined the effect of removal of carbon dioxide on the growth of this fungus in absence of various micro-nutrients. Under these conditions omission of zinc, copper or manganese from the culture medium reduced the growth of the fungus proportionately more when carbon dioxide was absent than when it was present. Steinberg concluded that the trace elements probably play a specific part in the utilization of carbon dioxide by *A. niger*, and compared this with the conclusion of Emerson and Lewis on the part they play in the utilization of carbon dioxide in green plants. Incidentally, Steinberg called attention to the significance of this similarity in regard to the question of the validity of the suggestion of Ruben and Kamen that carbon dioxide utilization by micro-organisms is essentially the same process as the dark or Blackman reaction in photosynthesis.

So far the functions of the trace elements in general have been considered. The specific functions of the individual trace elements will now be dealt with.

Manganese. The view has been very generally held that manganese is related to oxidation in the plant. Bertrand (1897), as we have seen, first called attention to the importance of manganese when he concluded that it was essential for the action of the oxidizing enzyme laccase. Later he maintained that it was essential for the action of oxidizing enzymes in general. Later work has not supported the view that manganese is generally involved in the action of oxidizing enzymes but has left no doubt that it is concerned in the respiration process. The oxidation of carbohydrate to carbon dioxide and water, which is the most usual type of respiration in plants, involves a long series of reactions, each one of which, with only one or two possible exceptions, is catalysed by an enzyme. These reactions are conveniently grouped into first glycolysis, a chain of reactions involving the phosphorylation of hexose sugar and its subsequent degradation to a three-carbon atom compound pyruvic acid,* $CH_3.CO$.

* All references to acids are to be regarded as including their salts; thus any reference to pyruvic acid applies equally well to pyruvates.

COOH, and secondly a cycle of reactions, the tricarboxylic acid or Krebs cycle, by which the pyruvic acid first condenses with oxaloacetic acid to give citric acid and carbon dioxide, after which the citric acid gives rise through a series of oxidations and decarboxylations to carbon dioxide and water with the re-formation of oxaloacetic acid.* It is now known that many of the reactions involved in glycolysis and the Krebs cycle are activated by manganese although with many of them activation can also be effected by some other ion, usually a divalent one, particularly magnesium, at any rate *in vitro*. For a few of the actions, however, manganese appears to be essential. This may be so with some enzymes of the Krebs cycle. One such is malic dehydrogenase which brings about the oxidation of malic acid $(COOH.CHOH.CH_2.COOH)$ to oxaloacetic acid $(COOH.CO.CH_2.COOH)$. An enzyme, oxaloacetic carboxylase, which effects the removal of carbon dioxide from oxaloacetic acid to produce pyruvic acid, and which appears to be generally associated with malic dehydrogenase in higher plants, also appears to require manganese for its action. It may be that both these actions are mediated by the same single enzyme but if this is so it appears to be distinct from a second enzyme known as the malic enzyme and shown to be widely distributed in plants and which also effects the production of pyruvic acid from malic acid. This enzyme also requires manganese for its action but *in vitro* it has been found that the manganese can be replaced by cobalt.

Another enzyme operating in the Krebs cycle which is likewise dependent on manganese which can be replaced by cobalt is oxalosuccinic decarboxylase which brings about the separation of carbon dioxide from oxalosuccinic acid $(COOH.CO.CH(COOH).CH_2.COOH)$ with the production of α-ketoglutaric acid $(COOH.CO.CH_2.CH_2.COOH)$. Yet Nason (1952) found the activity of this enzyme in extracts of tomato leaves deficient in manganese was twice that of extracts from normal leaves.

Among the enzymes concerned in the glycolytic stage of respiration the activation of which can be brought about by manganese or some other metallic ion are hexokinase which

---

* For a brief account of the series of reactions involved in respiration see W. Stiles and W. Leach, *Respiration in Plants* (fourth edition, London, 1960).

brings about the phosphorylation of hexose to produce hexose-6-phosphate, phosphoglucomutase which converts glucose-1-phosphate, obtained from the breakdown of starch, to glucose-6-phosphate, and enolase which catalyses the withdrawal of water from 2-phosphoglyceric acid to produce 2-phosphoenolpyruvic acid, the precursor of pyruvic acid. Of enzymes catalysing reactions in the Krebs cycle which are activated by manganese or some other metal there are the condensing enzyme by which citric acid is produced from oxaloacetic acid, isocitric dehydrogenase which catalyses the formation of oxalosuccinic acid from isocitric acid, and α-ketoglutaric oxidase (dehydrogenase or decarboxylase) which effects the oxidation of α-ketoglutaric acid to succinic acid with production of carbon dioxide. Although some other ions can replace manganese ions as activators of these reactions, manganese is generally the most effective and appears to be the metal of major importance in the tricarboxylic acid cycle. As with oxalosuccinic dehydrogenase Nason found the activity of isocitric dehydrogenase in extracts from manganese-deficient tomato leaves was twice that of extracts of normal leaves.

Another enzyme concerned in respiration which was found to require manganese, at any rate in soya beans, is carboxylase (pyruvic decarboxylase). This is the enzyme which brings about the removal of carbon dioxide from pyruvic acid with formation of acetaldehyde, and is the means by which carbon dioxide is produced in anaerobic respiration or fermentation.

Not only is manganese an activator of enzymes catalysing various stages in respiration; it has also been found to be concerned with enzymes playing a part in nitrogen metabolism. A flavoprotein enzyme which brings about the reduction of nitrite to hydroxylamine, a supposed early stage in the synthesis of amino acids and proteins, was found in leaves of soya bean by Nason, Abraham and Averbach (1954) and for its action manganese was found to be essential. Another enzyme requiring manganese, which was identified in both algae and higher plants by Stumpf and collaborators (1950, 1951) brings about a reaction between hydroxylamine and the amide glutamine:

$$COOH.CHNH_2.CH_2.CH_2.CONH_2 + NH_2OH \rightleftharpoons$$
$$COOH.CHNH_2.CH_2.CH_2CONHOH + NH_3$$

An enzyme requiring either manganese or magnesium and concerned in nitrogen metabolism was isolated from peas by Elliott (1953). This enzyme was found to catalyse two reactions, the first being the production of the amide glutamine from the amino acid glutamic acid and ammonia, the second the production of glutamine or glutamohydroxamic acid by the transfer of the γ-glutamyl group to ammonia or hydroxylamine respectively. But whereas magnesium was found to be more effective than manganese in effecting the first of these reactions, the reverse was the case with the second.

A number of enzymes concerned in the breakdown of peptides also require the presence of a metal. Generally activation of these peptidases can be effected by more than one metal and these generally include manganese. Among such are peptidases obtained from cabbage, spinach, barley and jackbean. For the action of one peptidase, prolidase, which catalyses the production of proline from glycylproline, manganese appears to be specifically necessary.

These observations on the activation of enzymes *in vitro* indicate that manganese plays an essential part in respiration and nitrogen metabolism. It has also been suggested that of the enzymes concerned in the Krebs cycle the malic enzyme and isocitric dehydrogenase may also operate in photosynthesis acting in the reverse direction so that malic acid is produced from pyruvic acid and isocitric acid from α-ketoglutaric acid.

While most of the oxidases, the enzymes which effect oxidation by means of molecular oxygen, contain either iron or copper, one has been reported by Humphreys (1955) as occurring in wheat embryos which brings about the oxidation of coenzyme 2 or TPN (triphosphopyridine nucleotide) in its reduced form (TPNH) and which requires manganese for its action. An enzyme system bringing about the oxidation of auxin (indoleacetic acid) which has been called indoleacetic oxidase has also been reported as requiring manganese for its action. According to Waygood, Oaks and MacLachlan this system is a complex one, the actual oxidation of the auxin being brought about by the catalytic action of trivalent manganese ions. The oxidation of dihydroxymaleic acid by a peroxidase described by Theorell and Swedin (1939) was also held by them to require manganese for its activation.

Some of the observed effects of manganese in plant activities could be expected from the relation of this element to enzyme actions. Thus Lundegårdh (1939) found the respiration of wheat roots as measured by oxygen absorption was increased by from 155 to 470 per cent by the addition of 0·00005M manganese chloride to the medium. In *Aspergillus niger* Chesters and Rolinson (1950) found that partial deficiency of manganese resulted in a reduction in the rate of oxidation of carbohydrate with concomitant increase in the production of organic acids. Shortage of other nutrients could, however, produce a similar result. Anderson and Evans (1956) found that the activities of isocitric dehydrogenase and the malic enzyme in extracts from plants of the dwarf bean (*Phaseolus vulgaris*) depended on the manganese supply. When nutrient solutions contained an excessive amount of manganese the activities of these enzymes was much greater than in extracts from plants grown in solutions containing a sufficiency, but not excess, of manganese, although the activity of these enzymes was not further reduced as a result of a deficiency of manganese in the culture solutions. As regards isocitric dehydrogenase it has already been noted that manganese can be replaced by magnesium. Possibly in the manganese-deficient plants the manganese content was yet sufficient for the functioning of the malic enzyme or perhaps some unknown metal might have replaced manganese as activator.

In regard to nitrogen metabolism Burström (1939) came to the conclusion that the assimilation of nitrate by whole wheat roots or by wheat root pulp did not take place in the absence of manganese and iron, but that when manganese was present in low concentration the assimilation of nitrate took place both with whole roots and root pulp, the maximum effect being produced with about 4 mg manganese per litre with whole roots and about 12·3 mg per litre with pulp. Without added manganese the addition of iron brought about feeble nitrogen assimilation of whole roots but not of pulp, and even this assimilation by whole roots was ascribed by Burström to its effect on respiration and ion uptake. His general conclusion was therefore that manganese, and not iron, directly catalysed nitrate assimilation.

The work of Jones, Shepardson and Peters (1949) also demonstrated the dependence of nitrate assimilation on manganese.

Soya bean plants were grown in nutrient solutions or in solutions of calcium nitrate, some with and some without manganese. The solutions were covered with a layer of oil in order to exclude oxygen. Under these conditions the plants supplied with manganese appeared normal and nitrite did not accumulate, whereas the leaves of plants not receiving manganese became yellow, the roots turned brown and large quantities of nitrite accumulated. The yellowing of the leaves was considered to be a symptom of nitrogen deficiency and the browning of the roots to be due to shortage of oxygen, while the accumulation of nitrite indicated a failure of its reduction, this leading to the nitrogen shortage in the leaves. This result was to be expected from the lack of activation of nitrite reductase which, as noted earlier, was demonstrated in soya bean leaves and shown to require manganese for its activation.

Observations by Heintze (1956) also indicated a disturbance in normal nitrogen relations in peas as a result of manganese deficiency. She found the amino acid content of peas exhibiting the manganese-deficiency symptom of marsh spot was higher than that of normal peas, while addition of zinc as zinc sulphate or molybdenum as sodium molybdenate both intensified the deficiency symptoms and increased the amino acid content of the peas. There is here no suggestion that shortage of manganese has inhibited the action of nitrite reductase and the result might be related to lack of activation of peptidase. But in the complex system of the cell there must be interactions between various substances which bring about changes not immediately predictable from actions observed in in vitro experiments; and this may be so here.

Several examples of this, as regards the effect of manganese on a number of enzyme actions, can be cited. For example, Bailey and McHargue (1944) examined the effect on the activity of a number of enzymes in extracts of lucerne (*Medicago sativa*) plants grown in water cultures supplied with manganese and some other micro-nutrients in different concentrations. Manganese was supplied in two concentrations, 1 and 2 p.p.m. Manganese at both levels of supply brought about a reduction in the activity of oxidases, catalase and sucrase, while with the weaker solution peroxidase activity was increased over that resulting with the control without added manganese, while with the stronger solution peroxidase activity was unaffected. With rice grown in water cultures containing manganese in concentrations ranging from 1 to 50 p.p.m. Pattanaik (1950b) found with the leaves a maximum activity of catalase when the manganese content of the medium was 5 p.p.m. while with the roots maximum activity of this enzyme was observed with manganese in a concentration of 1 p.p.m.

In an earlier chapter attention was called to the importance of the iron/manganese ratio in determining the incidence of symptoms of manganese deficiency or excess. A number of workers have seen in this an indication that a function of manganese in plants is to be found in its relation to the oxidation and reduction effected by iron salts. Thus Hopkins (1930), from observations on the growth of the unicellular green alga *Chlorella*, held that manganese brought about the reoxidation of iron after its reduction in the plant to the ferrous state; if the amount of manganese in the plant was deficient there resulted too high a proportion of ferrous iron, while if manganese was present in excess the reduction of ferric iron was prevented and the concentration of ferric iron was too high. In either condition the oxidation-reduction processes of the cell involving iron were disturbed.

This view of the relation between manganese and iron in the plant was dealt with in more precise terms by Shive (1941). This writer made two important points in this connexion, the first that the active functional iron in the tissues is in the ferrous state, the second that the oxidizing potential of manganese is higher than that of iron. Shive held that if iron is absorbed in the ferric state much of it is reduced in the plant by powerful reducing systems unless this is prevented by a counter-reactant. The manganese functions as such a counter-reactant, oxidizing ferrous to ferric iron which is precipitated, probably in organic complexes. Hence, if manganese is deficient in the plant, there will be an excess of active ferrous iron which induces chlorosis, a chlorosis due to iron toxicity. On the other hand, if the concentration of manganese is high the concentration of active ferrous iron is low, and if too low a chlorosis due to iron shortage will result. Thus it is necessary for healthy growth that the proportion of iron to manganese should lie within certain limits, and

Shive concluded that, for the species investigated by him, the ratio of active iron to active (soluble) manganese in the plant should lie between 1·5 and 2·5. This conclusion was derived from the results of experiments with soya bean carried out by Somers and Shive (1942) and described earlier (p. 120).

In further support of the oxidation-reduction hypothesis outlined above, Somers and Shive mentioned a series of tests carried out with maize seedlings in which cobalt was substituted for manganese. The oxidation potential of cobalt is higher than that of manganese, so it should, on the hypothesis, have a greater tendency than manganese to lessen the metabolic efficiency of iron by effecting the oxidation of the latter to the insoluble ferric state, and this, Somers and Shive state, was so.

The oxidation-reduction system hypothesized by Shive was questioned by Hewitt (1948 a, b) since it would not explain why some other metals such as copper can also induce iron deficiency and the resulting chlorosis, which might be due, as suggested by Hewitt and Sideris and Young (1949), to competition with and replacement of iron by another metal in actions normally leading to the formation of chlorophyll, the production of which is thereby inhibited.

Zinc. As far as is known at present zinc enters into considerably fewer enzyme systems than does manganese. The first enzyme to be recognized as containing zinc as an essential part of its molecule was carbonic anhydrase which catalyses the decomposition of carbonic acid to carbon dioxide and water: $H_2CO_3 \rightleftharpoons CO_2 + H_2O$. The presence of zinc in this enzyme obtained from animal sources was shown by Keilin and Mann in 1940. The presence of the enzyme in plants was first shown by Neish (1939) who demonstrated its presence in protoplasm and chloroplasts of wild red clover (*Trifolium pratense*) and the fern *Onoclea sensibilis*. Subsequent work has shown it to be widely distributed throughout the plant kingdom. According to Waygood and Clendenning (1950) carbonic anhydrase occurs in the cytoplasm of land plants and in the chloroplasts of aquatics.

According to Vallee, Hoch, Adelstein and Wacker (1956) another enzyme containing zinc or dependent on zinc for its action is triosephosphate dehydrogenase, which is concerned in the oxidation and further phosphorylation of phosphoglyceric aldehyde with the production of diphosphoglyceric acid, an essential step in the glycolysis of carbohydrate in the respiration and fermentation processes.

Zinc has been found to be an essential constituent of some enzymes present in fungi. Thus, according to Vallee and Hoch (1955) this element is an essential constituent of the alcohol dehydrogenase of yeast, and Nason, Kaplan and Colowick (1951) and Nason, Kaplan and Oldewurtel (1953) found zinc-deficient mycelia of *Neurospora crassa* were devoid of alcohol dehydrogenase activity. From this same fungus Medina and Nicholas (1957) obtained a hexokinase (see p. 149) for which the presence of zinc was essential. Zinc has been found to be among the metals which will activate aldolase, the enzyme catalysing the splitting of fructose diphosphate to two triosephosphates, and enolase (see p. 150) of yeast.

Zinc, like manganese, can affect the activity of enzymes for which its presence is not essential. Bailey and McHargue (1944) found the activity of peroxidase, catalase and oxidase in lucerne were all increased by the addition of zinc in concentrations ranging from 0·5 to 2 p.p.m. to the solutions in which the plants were growing, maximum activity resulting when the zinc concentration was 1 p.p.m. The effect of zinc on sucrase activity was the reverse, and minimum activity of this enzyme occurred with this same zinc concentration.

With a fumaric-acid-producing strain of the fungus *Rhizopus nigricans* Foster and Denison (1950) found that the carboxylase activity of the expressed juice of zinc-deficient mycelia was negligible, whereas the juice from mycelia with an adequate zinc supply possessed a high carboxylase activity. This difference was reflected in the intensities of fermentation by the respective mycelia, that of the plants with adequate zinc being from 7 to 10 times that of the zinc-deficient mycelia.

An effect of zinc on the production of the iron-containing cytochrome c in the smut *Ustilago sphaerogena* was recorded by Grimm and Allen (1954). As the concentration of zinc in the medium was increased up to 1 p.p.m. so was the production of cytochrome c when ammonium nitrogen was present.

If zinc is essential for the action of triosephosphate dehydrogenase, a stage in glycolysis, it may be expected that deficiency of zinc will involve a disturbance in the oxidation processes of the plant, although zinc is not a constituent of the actual oxidases. It has indeed been suggested that zinc is concerned in oxidations in the plant, but if the effects of zinc deficiency described below are related to suppression of the action of triosephosphate dehydrogenase, the connexion is quite unclear.

That zinc was concerned in oxidation-reduction reactions was the conclusion reached by Reed and Dufrénoy (1935) mainly as a result of their microscopical examination of mottled leaves of *Citrus*. They concluded that zinc was concerned with the functioning of sulphydryl compounds such as cysteine in their regulation of the oxidation-reduction potential within the cells. We have already noted that they found that the stromata of the chloroplasts in the palisade cells of such leaves were often rich in fat, while the vacuoles contained phenolic material and phytosterol or lecithin, which were absent from normal leaves. Reed and Dufrénoy interpreted these substances as suboxidized products of carbohydrates and proteins, and their presence suggested a disturbance in the oxidation-reduction potential within the leaf cells.

For the view that the maintenance of the oxidation-reduction potential at its normal level depends on sulphydryl compounds the following arguments were advanced. First, Hopkins and others have shown that such compounds appear to be present in all living cells and may control oxidation and reduction processes. Secondly, the oxidation of cysteine to cystine is catalysed by metals.* Thirdly, Giroud and Bulliard have shown that zinc has a specific effect in stabilizing the nitroprusside colour reaction of the sulphydryl (SH) group.† Reed and Dufrénoy also referred

* It is stated, however (cf. Meldrum, 1934), that the most active metals are iron, copper and manganese.

† This test, due to Mörner, consists in adding a 5 or 10 per cent solution of sodium nitroprusside rendered alkaline with ammonia to the liquid to be tested, and shaking. A violet colour, which soon fades, is produced if a cysteine peptide such as glutathione is present. Some other substances, including creatinine and acetone, give somewhat similar colours. According to Giroud and Bulliard the addition of salts of zinc gives a red colour much more stable than the violet colour of the reaction as usually produced. It appears to be quite specific for the sulphydryl group and is not given by either creatinine or acetone.

to Mazé's finding that roots of maize grown in a culture solution deficient in, though not completely free from, zinc, contained sulphides in the ash, indicating that the sulphur metabolism of the plant was adversely affected.

Chandler (1937) pointed out that zinc deficiency has its most serious effects in plants where carbohydrates have accumulated, and he suggested that zinc deficiency brings about inhibition of some process of carbohydrate transformation. This, as Reed (1938) suggested, may mean that zinc catalyses oxidation processes which in its absence may run the other way.

Following on their earlier work, Reed and Dufrénoy (1942) made a cytological study of catechol aggregates which arise in the vacuoles as a result of zinc deficiency. They considered that they form by the process called coacervation, in which disperse phase particles of the colloidal system constituting the vacuole become aggregated into spherical masses. This process was regarded as something more than a separation of phases, as the aggregates become surrounded by a precipitation membrane composed of orientated molecules of a phospholipid. Tests for oxidase showed (Dufrénoy and Reed, 1942) that these aggregates are not only centres of catechol derivatives, but also for catechol oxidase activity.

Reed and Dufrénoy concluded that the vacuolar sap contains both oxidizable phenolic compounds and catechol oxidase capable of catalysing the oxidation of these compounds. Normally this oxidation is prevented owing to the presence of hydrogen donators which may include the ascorbic-dehydroascorbic acid system, dihydroxymaleic acid, cysteine and glutathione. During the earlier part of the growing season the relatively high concentration of hydrogen donators in the cell protects the catechol compounds from oxidation and they remain dispersed throughout the vacuole. With the approach of senescence, or with nutrient deficiency such as a shortage of zinc, the oxidation-reduction equilibrium is disturbed and coacervation results, the process, according to Reed and Dufrénoy, being a 'simple consequence of a gradient in the distribution of cations and correlative distribution of polyphenol oxidase'. They further supposed that a difference in electrical potentials will exist between the aggregations and the surrounding medium, since the former are foci for catechol oxidase, and that there will therefore be a tendency for cations to move into the coacervate, and this in turn will greatly influence the intake of dissolved material by the cell and consequent derangement of metabolism.

A somewhat more precise suggestion of the way in which zinc, through its effect on oxidation-reduction systems, may affect growth, was put forward by Skoog (1940). Experiments made and described by this worker on tomatoes and sunflower grown in zinc-deficient culture solutions indicated a connexion between zinc and the production of auxin. Terminal buds and stems of such zinc-deficient plants appeared to contain no auxin or only a trace of it. Appreciable amounts were, however, found to be present in the leaves, although less than in leaves of control plants provided with an adequate supply of zinc. The visible symptoms of zinc deficiency only appeared after the decrease in auxin content, and if plants in an extreme state of zinc deficiency were supplied with zinc, the auxin content of these plants increased considerably in one to a few days, while growth was resumed after the passage of several more days. These observations suggested that zinc was necessary for the maintenance of a normal auxin content.

Skoog also placed sections of stems from zinc-deficient and control plants, from which auxin had previously been removed, on agar blocks containing a known concentration of indole-3-acetic acid, and found that always more of this growth hormone was inactivated in the blocks in contact with tissue from zinc-deficient plants than in the blocks in contact with control tissue. This suggested that deficiency of zinc brought about excessive destruction of auxin, an effect attributed to an increased oxidative activity of the tissues, since these displayed increased capacity to oxidize benzidine in presence of hydrogen peroxide. That zinc functions as a catalyst in relation to oxidation-reduction processes in the cell is thus again indicated, while its relationship to auxin maintenance suggests why zinc deficiency may lead to retardation or cessation of growth.

Later work by Tsui (1948) indicated that the effect of zinc deficiency on auxin was to prevent its formation rather than to induce its destruction. Tsui found that the auxin content of tomato plants grown in culture solutions supplied with zinc increased progressively with the age of the plant but in cultures without zinc the auxin content decreased with age and this occurred before there was any falling off in growth rate or any visual symptoms of zinc deficiency. Dried material was found to contain two kinds of bound auxin, one acid stable and alkaline labile, the other the reverse. In plants with an adequate zinc supply the former kind increased with the growth of the plant while the alkaline stable kind remained practically constant. In zinc-deficient plants, on the other hand, both kinds decreased as the plants grew older. Addition of zinc to the culture solution resulted within two days in an increase in the content of free auxin and enzyme-digestible auxin, which was followed after another two days by increase in the quantity of the bound auxins. Indoleacetic acid is closely related chemically to tryptophane and this amino acid is commonly regarded as a precursor of auxin. It was therefore significant that Tsui found the tryptophane content of zinc-deficient plants considerably less than that of plants adequately supplied with zinc, and that addition of zinc to a previously zinc-deficient culture solution was followed in three days by increase in the tryptophane content of the plants. It was therefore concluded that zinc was essential for the synthesis of tryptophane and so only indirectly for the synthesis of auxin. That zinc was not necessary for the formation of auxin from tryptophane was indicated by the results of experiments in which disks of leaf tissue were provided with tryptophane. The production of active auxin from this was then the same in both normal and zinc-deficient tissue.

In the fungus *Neurospora crassa* also it was found by Nason and his collaborators (1950, 1951, 1953) that zinc was required for the production of tryptophane. Here it was supposed that absence of zinc prevented the formation of tryptophane from indole and serine by the action of an enzyme which required zinc for its formation but not for its action since addition of zinc to the medium or to extracts of the fungus did not activate the enzyme.

Copper. Several plant enzymes are known which contain copper. These are polyphenol or catechol oxidase, tyrosinase, laccase and ascorbic acid oxidase. These are all enzymes which bring about the oxidation of organic compounds by means of molecular oxygen.

The first enzyme shown to contain copper was polyphenol oxidase, an enzyme or group of similar enzymes widely distributed in plants. The presence of copper in the enzyme was shown by Kubowitz (1937) for the oxidase present in potato tuber and by Keilin and Mann (1938) for the similar oxidase present in the cultivated mushroom (*Agaricus campestris*). The enzyme, in fact, appears to be a copper–protein compound containing not less than 0·30 per cent of copper.

Catechol or polyphenol oxidase catalyses the oxidation of compounds with the orthodihydroxyphenol grouping such as catechol and pyrogallol. In the presence of the enzyme and a low concentration of catechol a number of other substances such as guaiacum, benzidine and ascorbic acid are also oxidized which are not oxidized by the pure enzyme alone. These further oxidations may require the presence of some substance such as orthoquinone, produced by the oxidation of catechol in the primary reaction, so the presence in plant tissues of a substance either of the catechol or orthoquinone type along with the oxidase would make possible the oxidation of a wide range of phenolic compounds.

Tyrosinase is the name given to an enzyme which has been known for sixty years to catalyse the oxidation of the phenolic amino acid tyrosine and some other phenolic compounds. As these include monohydric phenols the enzyme has also been called monophenol oxidase. There is some doubt whether tyrosinase is distinct from polyphenol oxidase, it having been suggested that the oxidation of tyrosine is brought about by a secondary reaction resulting from polyphenol oxidase activity as, for example, by orthoquinones produced in that action. That tyrosinase contains copper was shown by Nelson and Dawson (1944).

Laccase is an enzyme which catalyses the oxidation of diphenols uroshiol and laccol which are present in the latex of the lacquer trees *Rhus succedanea* and *R. nucifera*. As mentioned earlier G. Bertrand considered that this enzyme was activated by manganese but Keilin and Mann (1939, 1940b) found it to be, like polyphenol oxidase, a copper–protein compound containing about 0·2 per cent of copper.

Ascorbic acid oxidase, which catalyses the oxidation of ascorbic acid, is widely distributed in plants. That it is a copper–

162   THE FUNCTIONS OF

protein compound was shown by Lovett-Janison and Nelson (1940) and Meiklejohn and Stewart (1941). According to Dunn and Dawson (1951) the copper content is about 0·26 per cent.

Two of these copper enzymes, polyphenol oxidase and ascorbic acid oxidase, and a possible third, tyrosinase, have been held to be terminal oxidases in respiration; that is, they catalyse the actual oxidation in the last stage of aerobic respiration in which hydrogen is removed from reduced coenzymes produced during the Krebs cycle to give the water which is one of the final products of respiration. Materials in which polyphenol oxidase has been thought to act as a terminal oxidase are potato tubers and apples, while among materials in which it has been suggested that ascorbic acid oxidase is a terminal oxidase are barley seedlings, apple fruits and swedes. To what extent these copper enzymes are terminal oxidases has been a matter of some controversy, the more general opinion being that the most usual terminal oxidase is a cytochrome system which contains iron, although it may be, as James (1953) concluded was the case for barley, that as the plant grows older cytochrome oxidase may be replaced by ascorbic acid oxidase as the terminal oxidase of respiration.

Catechol oxidase from *Atropa belladonna* would appear also to catalyse the oxidation of amino acids and secondary amines, according to James, Roberts, Beevers and de Kock (1948) and Beevers and James (1948). The actual oxidation would appear to be effected by the orthoquinones produced in the oxidation of polyphenols.

In line with the presence of copper in these oxidases is the finding of Bailey and McHargue (1944) that the oxidase activity of tomato leaves and fruits of plants grown in water cultures containing copper in concentrations ranging from 0·0 to 0·1 p.p.m. increased with increase in the copper concentration.

Like manganese copper can affect the activity of enzymes not depending on copper for their activation. Bailey and McHargue found with the leaves of tomato plants that the activity of sucrase, catalase and peroxidase was greater when the concentration of copper in the culture solution was 0·01 p.p.m. than when it was higher. With the fruits sucrase activity increased with increase in copper concentration over the range 0·01 to 0·1 p.p.m. while catalase and peroxidase activity decreased with increase in copper concentration over this range, the decrease being greater with catalase than with peroxidase.

There is evidence that copper is concerned with the oxidation of iron in the plant. Erkama (1947, 1950), who determined the copper, iron and manganese contents of fifteen different species of flowering plants, decided that the ratio of copper to iron varied less than did that of manganese to iron over the range of plants examined, the ratio of copper to iron ranging from 0·033 to 0·181 while that of manganese to iron ranged from 0·31 to 9·15. Erkama also found that copper was strongly bound to the protoplasm while manganese appeared to dialyse easily from the cell. With peas he found that copper-deficient plants grown in sterile water cultures contained less iron than those supplied with copper. This contrasted with the effect of manganese deficiency which had the effect of increasing the iron content of the plants. Erkama considered that copper in the protoplasm oxidized ferrous iron to the insoluble ferric state so that it ceased to be active, playing much the same part in relation to iron as manganese played in the vacuole.

A relationship between copper and iron was also reported by Brown and Holmes (1955), but different plants were found to exhibit different behaviour in this respect. In maize, deficiency of copper resulted in accumulation of iron throughout the plant, the addition of copper to the soil reducing the uptake of iron. This result contrasts with that recorded by Erkama for peas. Differences between different plants in the effect of copper on iron metabolism were also recorded by Brown, Holmes and Specht (1955), who found that with increasing phosphate supply together with copper the absorption and utilization of iron in a variety of soya bean, PI–54619–5–1, decreased while wheat and another variety of soya bean, Hawkeye, were not affected in this way. The effect of copper on a number of plants examined indicated that the yield of those susceptible to iron chlorosis, namely, rice and the PI–54619–5–1 variety of soya bean, was reduced as the result of increasing additions of copper to the soil, while with plants not so susceptible to iron chlorosis, wheat and the Hawkeye variety of soya bean, this was not so.

Some effects of copper in furthering oxidation in bacteria and fungi were reported by Mulder (1950). The oxidation of ethyl

alcohol to acetic acid by *Acetobacter aceti* was much increased
by the addition of 0·5 μg of copper to 50 ml of culture solution,
and the formation of the black pigment in the spores of
Aspergillus niger and in cultures of *Azotobacter chroococcum* was
also found to require copper (cf. Table 6, p. 44).

Molybdenum. An enzyme which contains molybdenum is
nitrate reductase which catalyses the reduction of nitrate to
nitrite. That the activity of this enzyme in the fungi *Neurospora
crassa* and *Aspergillus niger* was dependent on molybdenum was
demonstrated by Nicholas, Nason and McElroy in 1953, and in
1955 Nicholas and Nason found that the action of nitrate reductase in soya bean was similarly dependent on molybdenum.
The presumption is that the first stage in the assimilation of
nitrate by plants, its reduction to nitrite, is dependent on the
presence of molybdenum.

Recognition of the fact that the function of molybdenum was
connected with nitrogen metabolism had been growing for some
time. As early as 1930 Bortels had found that the fixation of
nitrogen by the nitrogen-fixing bacterium *Azotobacter chroococcum* was stimulated by molybdenum. In 1936 Steinberg,
using nitrate as source of nitrogen for *Aspergillus niger*, found
that molybdenum was essential for the growth of this fungus but
later (1937) he found that when an ammonium salt was used as
nitrogen source the molybdenum requirement was much less.
Mulder (1948, 1950) also found that when this fungus was grown
with ammonium compounds as source of nitrogen the molybdenum requirement for normal growth was much less than when
nitrate was the nitrogen source, and he concluded that molybdenum was probably required as a catalyst in nitrate reduction.
In later work Mulder, Boxma and van Veen (1959) found the
nitrate-reducing power, as measured by nitrite production, of
coarsely cut fragments of plants grown in molybdenum-deficient
soil was very low.

Also in line with the necessity of molybdenum for nitrate
reduction was the finding of Hewitt and Jones (1947) that in
higher plants shortage of molybdenum led to accumulation of
nitrate in the leaves and a reduction in protein synthesis. Accumulation of nitrate in the leaves of tobacco was found by
Steinberg, Specht and Roller (1955) to result from shortage of

any trace element except copper, but was greatest as a consequence of molybdenum deficiency.

In experiments with cauliflower plants grown in sand cultures which were supplied with molybdenum in concentrations of 0·0055, 0·55 and 55 p.p.m. Hewitt, Jones and Williams (1949) found the content of a number of amino acids, alanine, aspartic acid, glutamic acid, proline and particularly arginine, decreased with decrease in the level of molybdenum supply. A decrease in the production of the amides asparagine and glutamine also occurred.

This connexion between molybdenum supply and the formation of amino acids in plants of cauliflower, spinach and tomato supplied with nitrate as source of nitrogen was later demonstrated by Mulder, Bakema and van Veen (1959). With spinach marked increases in the content of glutamic acid and glutamine were observed only two hours after the application of molybdenum to the soil in which molybdenum-deficient plants were growing, while slight increases were observed in the contents of aspartic acid and alanine.

The observation of Mulder that the sensitivity to molybdenum deficiency depended on the source of nitrogen supplied to the fungus *Aspergillus niger* was extended by Argawala (1952) in an investigation of the molybdenum requirements of cauliflower plants supplied with nitrogen in a number of different sources, which were nitrate with citric acid, nitrate with and without calcium carbonate, ammonium sulphate with calcium carbonate, ammonium nitrate, ammonium citrate, urea and nitrite. Molybdenum was supplied in four concentrations, namely, 0·00001, 0·00005, 0·5 and 50 p.p.m. Of these the first two were considered to be low enough to produce deficiency while the last was high enough to produce symptoms of molybdenum excess. The result was that the whiptail condition symptomatic of molybdenum deficiency developed whatever the source of nitrogen, with the one exception of nitrite, with molybdenum supplied at the lowest level. However, when supplied with ammonium sulphate and calcium carbonate visible symptoms of molybdenum deficiency did not appear when molybdenum was supplied at the still low level of 0·00005 p.p.m.

Thus it would appear that in cauliflower the molybdenum

requirement depends on the source of nitrogen and this was also shown to be so for tomatoes by Hewitt and McCready (1954) where the effects of molybdenum deficiency were examined with nitrogen supplied in a number of forms including nitrate, nitrite, ammonium sulphate, ammonium nitrate, ammonium nitrite, urea and glutamic acid. The visual symptom of molybdenum deficiency, chlorosis, was only observed when the nitrogen was supplied as nitrate, and this included ammonium nitrate. With ammonium sulphate there was, on the other hand, a definite increase in chlorophyll content. An internal effect of molybdenum deficiency, namely, a decrease in the content of both ascorbic acid and dehydroxyascorbic acid, was observed whatever the source of nitrogen, but the extent of the decrease depended on the form of nitrogen supplied, being greatest with nitrate and least with urea and nitrite.

It would thus appear that although it is well established that through its presence in nitrate reductase molybdenum is essential for the assimilation of nitrate by plants, yet since it is still required when nitrogen in a reduced form as in an ammonium salt is provided, it can be concluded that the reduction of nitrate to nitrite is not the only process for which it is essential. Further evidence of this was provided by the finding of Vanselow and Datta (1949) that symptoms of molybdenum deficiency appeared in lemon cuttings deprived of this element, even when the nitrogen was supplied as ammonium salt.

Reference has been made earlier to the reduction in the amount of ascorbic acid in tomato plants resulting from a shortage of molybdenum whatever the nitrogen source. Such a reduction in ascorbic acid content in molybdenum-deficient plants was also observed with alsike clover (*Trifolium hybridum*), barley, sugar beet and varieties of *Brassica* by Hewitt, Agarwala and Jones (1950).

Some other effects of molybdenum deficiency in certain species of plants have been noted. Argawala (1952) found that in cauliflower plants exhibiting the whiptail condition the cells appeared not joined together, with the middle lamella almost invisible, suggesting a possible abnormal metabolism of pectic substances. In tobacco Steinberg and Jeffrey (1956) found that deficiency of molybdenum brought about a decrease in the nicotine content but this effect was also induced by shortage of iron, copper, manganese and zinc, but only excess of molybdenum or zinc brought about increase in nicotine content, excess of the other elements mentioned bringing about decrease in nicotine content. Increased content of nicotine was always accompanied by excessive branching of the root system.

One plant in which it has been reported that while molybdenum is necessary if nitrate is the source of nitrogen, it is not essential if nitrogen is supplied as ammonium salt is, according to Ichioka and Arnon (1955), the alga *Scenedesmus*.

Reference has been made earlier to the observation of Bortels that molybdenum was found to be required for the fixation of nitrogen by the nitrogen-fixing organism *Azotobacter chroococcum*. Mulder (1950) examined the growth of this organism in nutrient solutions supplied with nitrogen as nitrate or as ammonium sulphate or when gaseous nitrogen was supplied, in relation to molybdenum supply. It was found that about ten times as much molybdenum was required when gaseous nitrogen was the nitrogen source as when nitrate was used, while there was no response to molybdenum when the source of nitrogen was ammonium sulphate. According to Horner, Burk, Allison and Sherman (1942) molybdenum could be partially replaced by vanadium in nitrogen fixation by *Azotobacter chroococcum*, while according to Nicholas (1958) tungsten can also partially replace molybdenum. Nicholas has also reported that nitrogen fixation by *Azotobacter vinelandii* and *Clostridium pasteurianum* was reduced to about 40 per cent by the omission of molybdenum from an otherwise complete nutrient solution. There would thus appear to be no doubt that molybdenum is essential for the fixation of nitrogen by nitrogen-fixing bacteria, but whether this involves another molybdenum-containing enzyme is at present not known.

Since molybdenum is concerned in both the reduction of nitrate by higher plants and in the fixation of elemental nitrogen by nitrogen-fixing bacteria, it would appear to be doubly concerned in nitrogen assimilation by members of the Leguminosae. Work by Anderson and Thomas (1946) and Anderson and Oertel (1946) indicated that addition of molybdenum to the soil had the effect of reducing the number of nodules on the roots of the leguminous plants lucerne (*Medicago sativa*) and subterraneum clover (*Trifolium subterraneum*) although Anderson and Spenser (1950) found with the latter plant that the nodules tended to be larger as a result of application of molybdenum. From work with white clover (*Trifolium repens*) on the interrelationships between molybdenum and calcium, Walker, Adams and Orchiston (1955) concluded that calcium and a sufficiently high pH were necessary for rapid nodulation and that molybdenum was concerned with the fixation of nitrogen by the nodule bacteria.

Mulder, Bakema and van Veen (1959) also stated that in absence of an adequate supply of molybdenum leguminous plants produce nodules but that these cannot fix nitrogen. Hence in molybdenum-deficient plants of lucerne, for example, the free amino acid content is very low, but some hours after the application of molybdenum to the soil the content of free glutamic acid in the nodules was found to have increased considerably while that of some other amino acids, alanine, aspartic acid and amino-butyric acid, rose slightly.

Boron. It has earlier been pointed out that the effects on plants of a deficiency of boron are generally of quite a different kind from those produced by other micro-nutrient deficiencies and indeed this element differs from the other four generally recognized trace elements in that it is a non-metal and does not appear to form part of any enzyme system. Nevertheless, it has been claimed that boron may affect the activity of a number of enzymes. Bailey and McHargue (1944) working with lucerne grown in culture solutions containing boron in concentrations ranging from 0·1 to 1 p.p.m. found that the activity of each of the enzymes catalase, peroxidase, oxidase and sucrase was un-affected by boron supplied in the lowest concentration of the range, but that as the concentration was raised above this the activity of all these enzymes increased, the effect on oxidase being the least and on sucrase the greatest. An increase in catalase activity as a result of adding boron in a concentration of 1 p.p.m. to the culture medium of rice was found by Pattanaik (1950a) and for sugar beet by Yakovleva (1947). Increased activity of one oxidase, tyrosinase, as a result of boron deficiency was reported by Klein (1951) for tomato and by Steinbeck (1951) for potato. Observations by MacVicar and Burris (1948) also suggested that boron had an inhibiting effect on some oxidases of tomato and soya bean for they found homogenates from leaves of boron-deficient plants of these species took up more oxygen than did similar preparations from plants adequately supplied with boron. Cell-free sap from the leaves of boron-deficient tomato and soya bean plants oxidized dihydroxyphenylalanine more rapidly than sap from the leaves of plants supplied with boron, suggesting a possible inhibiting action of boron on polyphenol oxidase. The oxidation of glycollic acid and lactic acid by cell-free sap from boron-deficient tobacco leaves was likewise inhibited by boron. Winfield (1945), on the other hand, decided that potato tyrosinase action was not affected by boron. The results concerning the effect of boron on oxidase action are thus conflicting. No doubt the results obtained also depended partly on other conditions as well as on the nature of the enzyme.

Quite a number of functions have been ascribed to boron in higher plants. These include effects on the water relations of plants, including the water relations of the protoplasm, water absorption by the plant and transpiration, a favourable influence on the absorption of cations and a retarding influence on the absorption of anions, a favourable influence on the absorption of calcium, a part in the formation of pectic substances in the cell wall, and an essential part in carbohydrate and nitrogen metabolism.

As regards the supposed effect of boron on the water relations of the protoplasm, it has already been pointed out that a very general symptom of boron deficiency, and often one of the earliest internal symptoms, is an enlargement of thin-walled cells. In this connexion some observations made by Schmucker (1933, 1935) on the germination of the pollen grains of tropical water lilies and other species in presence and absence of boron are of interest. He found that when the pollen was placed in a drop of nectar from the flower, germination was normal, but when placed in a drop of sugar solution of the same concentration, the grains either failed to germinate or the pollen tubes produced quickly burst. But the pollen tube remained intact and its growth continued if the sugar solution contained 0·001 or 0·01 per cent of boric acid. From such observations Schmucker not only concluded that boron influenced the absorption of water by the protoplasm but that along with sugar it played some part in the formation of pectin in the cell wall. With regard to the latter supposition we have already noted that one of the most general features of boron deficiency is the breakdown of parenchymatous and other thin cell walls, especially those of the apical meristems in which pectic substances are, by some, supposed to predominate owing to the relative importance of the middle lamella in such cell walls. Other features of boron deficiency, such as discoloration of cell walls and the brittleness of petioles and leaf laminae, might also be held to support this view. On the other hand, Dennis (1937) has pointed out that discoloration of the cell wall is not confined to cases of boron deficiency, and he suggests that the effect of boron deficiency on the cell wall may be part of a more far-reaching effect of this deficiency on carbohydrate metabolism.

The effect of boron deficiency on transpiration of the dwarf bean (*Phaseolus vulgaris*) growing in sand cultures was examined by Baker, Gauch and Duggar (1956), who found that the rate of transpiration from the leaves of boron-deficient plants was only 30 per cent of that from the leaves of plants adequately supplied with boron. Three factors were considered to bring about this result. In the first place the leaves of boron-deficient plants had a higher osmotic pressure and a higher content of polysaccharide, both factors tending to retention of water; secondly, deficiency of boron induced a lower rate of water absorption and thirdly, there was a higher proportion of non-functional stomata on the leaves of boron-deficient plants than on normal leaves.

The view that boron as boric acid increases the intake of cations and decreases that of anions was put forward by Rehm (1937) as a result of experiments with *Impatiens balsamina* grown in water culture. The cultures were provided either with single salts or complete nutrient solutions with and without boron. The intake of the different ions was determined by analysis of the solution with the results shown in Tables 27 and 28.

These data certainly indicate an increase of cation absorption and a relative or absolute decrease of anion absorption as a result of the presence of boron. At present it would seem to be extremely doubtful whether such an effect of boron can be accepted as a general one.

TABLE 27. *Absorption by* Impatiens balsamina *of various ions from solutions of single salts in presence and absence of boron.* (Data from Rehm)

Salt	Concentration	Concentration of boron, when present (mg per litre)	Ratio of uptake of ions in presence and in absence of boron	
			Cation	Anion
K$_2$SO$_4$	0·001N	0·5	1·378	1·05
KCl	0·001N	0·5	1·718	1·07
KH$_2$PO$_4$	0·002N	0·5	20·82	14·56
KH$_2$PO$_4$	0·002N	10	95·00	27·10
CaCl$_2$	0·002N	0·5	1·013	0·789
MgSO$_4$	0·002N	0·5	3·022	1·244
MgSO$_4$	0·002N	10	3·055	1·065
NH$_4$NO$_3$	0·002N	1	1·038	0·872

TABLE 28. *Absorption by* Impatiens balsamina *of various ions from solutions containing the major plant nutrients with and without boron.* (Data from Rehm)

Ion	Ratio of absorption of ion with boron present (0·5 mg/litre) to absorption with boron absent	
	Nitrogen supplied as Ca(NO)$_2$	Nitrogen supplied as NH$_4$NO$_3$
NH$_4$	—	1·61
K	1·215	4·46
Ca	1·24	1·09
Mg	1·045	1·03
NO$_3$	1·834	1·05
Cl	0·788	1·127
SO$_4$	0·721	1·022
H$_2$PO$_4$	0·74	0·17

A quantity of evidence has been presented by different workers indicating that boron may be associated with the utilization of calcium. This was suggested by the results obtained in a series of water-culture experiments with *Vicia faba* reported by Brenchley and Warington (1927). Cultures without boron and with a very small supply of calcium as calcium sulphate (0·025 g per litre) made poor growth and developed very small blackened leaves, finally turning black at the stem apices and withering backwards from these. It will be noted that these are the symptoms of calcium shortage and not of boron deficiency (see p. 83). The blackening is attributed to the toxic effect of other nutrient salts, a toxicity which is normally counteracted by calcium, and the plants probably died before even the calcium in the seed was used up. In cultures containing the same supply of calcium, but also containing boric acid, development was very much better, and, although there was some blackening of the leaves, the plants grew fairly tall, the stems did not blacken and normal flower buds developed. Brenchley and Warington interpreted these results as indicating that without boron the plants were unable to absorb or utilize sufficient calcium to prevent poisoning by the other nutrient salts, while when boron is present the latter enables the plant either to absorb calcium more rapidly or to utilize it more readily so that the toxic effect of the other nutrients is antagonized. It may be noted that in cultures without boron, as the supply of calcium is increased the symptoms of calcium shortage become less marked until with 0·1 g of calcium sulphate per litre they disappear and the plants show typical symptoms of boron deficiency.

Later, Warington (1934), by actual determination of the amount of calcium absorbed by plants growing in culture solutions, showed that the presence of boron does indeed result in a very considerable increase in the amount of calcium absorbed by plants of *Vicia faba*. The actual values she obtained are summarized in Table 29.

TABLE 29. *Absorption of calcium by* Vicia faba *from nutrient solutions with and without boron.* (Data from Warington)

Age of plant in weeks	Total calcium absorbed (mg per plant)	
	With boron	Without boron
1930 series: Solutions not renewed		
2	2·2	2·6
3	8·3	5·1
4	15·1	7·1
5	25·1	6·6
1931 series: Solutions not renewed		
2	2·2	2·2
3	6·2	3·5
4	13·8	5·5
5	17·1	6·0
Solutions renewed fortnightly		
5	17·8	10·1
9	50·9	15·6
Solution renewed weekly		
9	54·83	26·65

Swanback (1939), to whose work on absorption of cations by tobacco reference has already been made, also concluded from analyses of tobacco plants supplied with calcium and potassium at different levels with and without boron, that the latter element aids the absorption and utilization of calcium.

Results comparable with those of Miss Warington have been obtained with soya bean by Minarik and Shive (1939). Plants were grown in sand cultures supplied with the usual major nutrients and manganese. Boron was supplied to the different cultures in concentrations varying from 0 to 10 p.p.m. As the results summarized in Table 30 show, the boron supply definitely influenced the amount of calcium which accumulated in the leaves, and indeed the effect of boron on the growth of the plants was parallel with its effect on calcium uptake.

TABLE 30. *Absorption of calcium by* Glycine max *from media containing various amounts of boron.* (Data from Minarik and Shive)

Concentration of boron in medium (p.p.m.)	Average fresh weight of leaves per plant (g)	Calcium in leaves (mg per g of fresh weight)
0·0	4·7	2·6
0·001	11·0	2·5
0·0025	15·8	2·6
0·005	16·2	2·9
0·010	24·0	3·0
0·025	32·8	3·5
0·05	38·7	4·5
0·1	32·6	4·0
0·25	28·4	3·9
0·5	28·1	3·4
1·0	31·5	3·5
2·5	13·6	4·2
5·0	12·8	3·6
10·0	0·8	2·0

On the other hand, Holley and Dulin (1937), working with cotton, could find no indication of a relationship between boron and calcium, and Morris (1938) found no difference in calcium content in normal and boron-deficient oranges, while Talibli (1935) actually found that addition of boron to the medium on which flax was growing brought about a reduction in the calcium content of the flax straw.

Work by Marsh and Shive (1941) on maize appeared to throw some light on the apparent divergence between the results obtained up to that time with these various species. Plants were grown in sand cultures and for the first week were supplied with a nutrient solution containing all necessary elements except boron, this being omitted so that any in the seed or in external sources should be exhausted. During the second week all the cultures were supplied with a complete culture solution which included 0·25 p.p.m. of boron as boric acid. From the beginning of the third week two series of cultures receiving no calcium, and four supplied with a nutrient solution containing 170 p.p.m. of calcium, received different amounts of boron. These treatments were continued for 10 days, at the end of which time the dry weights of the shoots, the total calcium and boron content and the contents of soluble calcium and soluble boron of the shoots were determined. The various treatments and the results obtained from them are indicated in Table 31.

TABLE 31. *Calcium and boron content of maize supplied with different amounts of boron.* (Data from Marsh and Shive)

Concentration of boron supplied (p.p.m.)	Dry weight of shoot per plant (g)	Total Ca per g of dry matter (mg)	Soluble Ca per g of dry matter (mg)	Total B per g of dry matter (mg)	Soluble B per g of dry matter (mg)
		No calcium supplied in nutrient solution			
0·0	1·50	3·0	0·3	0·001	0·0005
0·25	2·60	3·0	1·0	0·008	0·0069
		170 p.p.m. calcium supplied in nutrient solution			
0·0	2·85	7·6	2·1	0·002	0·0015
0·1	5·40	7·7	2·4	0·005	0·0042
0·25	5·30	8·0	2·8	0·008	0·0070
5·0	4·37	7·7	4·2	0·025	0·0232

Inspection of these results shows that the *total* calcium content of the shoots is independent of the amount of boron supplied; on the other hand, the *soluble* calcium content runs parallel with both the soluble boron content and the total boron content of the plant and also with the boron content of the medium. It was therefore concluded that the soluble calcium content is determined by the boron content, a large proportion of which is in a soluble form, and which is itself determined by the boron content of the medium.

There is thus, according to Shive (1941), a difference between dicotyledons as exemplified by *Vicia faba* (and presumably *Glycine max*) and monocotyledons as exemplified by *Zea mais*, in that in the former the calcium and boron contents are generally much higher than in the latter, but in the dicotyledons studied only a small fraction of the boron is soluble, whereas in monocotyledons practically all the boron remains in solution. In both groups the soluble calcium is directly related to the soluble boron which is itself determined by the total boron, which in its turn is determined by the concentration of the boron in the medium. This much smaller proportion of soluble boron in dicotyledons explains why the boron requirement of these plants is so much higher (5–10 times) than that of monocotyledons.

Reference has been made earlier (p. 136) to the investigation of boron/calcium relationships in another monocotyledon *Setaria italica*. As regards absorption of calcium, as the results presented in Table 32 show, the total calcium in the shoots was roughly independent of the level of boron supplied with lower boron concentrations but showed a definite increase in total calcium with the higher boron concentrations. As with maize the *soluble* calcium content ran parallel with the total and soluble boron in the plants and with the boron content of the medium. Comparable results were obtained with higher contents of calcium in the medium.

TABLE 32. *Calcium and boron content of shoots of Siberian millet supplied with different amounts of boron.* (Data from McIlrath and de Bruyn)

Concentration of boron supplied (p.p.m.)	Dry weight of shoot per plant (g)	Total Ca per g of dry matter (mg)	Soluble Ca per g of dry matter (mg)	Total B per g of dry matter (mg)	Soluble B per g of dry matter (mg)
40 p.p.m. calcium supplied in medium					
0·0	1·53	2·61	0·058	0·029	0·022
0·05	1·55	2·41	0·093	0·020	0·009
0·50	1·55	2·74	0·260	0·043	0·026
5·0	1·31	3·79	0·260	0·262	0·231
50·0	0·34	4·16	1·165	1·116	1·100
160 p.p.m. calcium supplied in medium					
0·0	1·71	5·08	0·126	0·040	0·030
0·05	1·51	6·25	0·146	0·026	0·011
0·50	1·97	5·40	0·135	0·045	0·030
5·0	1·39	6·68	0·188	0·278	0·249
50·0	0·40	9·03	4·639	0·879	0·814

While, then, there is a quite considerable amount of evidence of a relationship between boron and calcium, and also between boron and potassium, in plant nutrition, the results obtained with different species are insufficient for any generalization as to the nature of this relationship at present.

There can be little doubt that boron plays some part in carbohydrate metabolism. Reference has already been made to the conclusion of Schmucker (p. 170), derived from experiments with germinating pollen, that boron played some part in the formation of pectin in the cell wall.

Microchemical tests made by Marsh and Shive (1941) on the apical meristem of maize plants supplied with different quantities of boron suggested a relation between boron and the pectin and fat contents of these tissues. Thus with a supply of 170 p.p.m. of calcium without boron, tests for pectin in the cytoplasm were positive and for fats negative, but with 5 p.p.m. of boron tests for pectin in the cytoplasm were negative and for fats positive. With an intermediate supply of boron (0·1 and 0·25 p.p.m.) which was found optimal for growth (cf. Table 31), tests for both pectin and fat were positive. It is suggested therefore that boron plays some part in carbohydrate and fat metabolism.

That boron is connected with carbohydrate metabolism was indicated in the work of Johnston and Dore (1929), who found that in tomato plants suffering from boron deficiency there was a marked accumulation of sugars in the leaves and a corresponding reduction of the sugar content of the stems, indicating some considerable reduction below normal in the translocation of carbohydrates.

That its effect on translocation was a leading factor in the importance of boron in carbohydrate metabolism was strongly advocated by Gauch and Duggar (1953, 1954). Their view was that the borate ion reacted with hydroxyl-rich substances such as sugars and sugar alcohols to produce complexes which were translocated more rapidly across membranes than undissociated free sugars. To account for this they suggested as a likely explanation that boron was a constituent of the membranes through which the sugar passed and that sugar combined with borate in a reversible reaction which would result in the rapid transference of sugar across the membrane. Shortage of boron would thus reduce the supply of carbohydrate to the growing regions and lead to failure of growth and finally to the breakdown of the tissues so often reported to occur in those tissues as a result of boron deficiency.

Data indicating a relation between boron and translocation of sugars in the tomato plant were presented by Sisler, Duggar and Gauch (1956). Plants growing in sand culture were supplied with boron at various levels and a solution of sucrose containing radioactive carbon ^{14}C applied to the leaf tips. The subsequent distribution of sugar was determined by measuring the radioactivity throughout the plants. In boron-deficient plants exhibiting deficiency symptoms the uptake and translocation of the sugar were less than with plants supplied with boron, and before the appearance of visible symptoms it was found that the rate of translocation of sugar in plants supplied with 50 p.p.m. of boron in the medium was more rapid than in plants without this supply. The same workers also examined the translocation of sugar produced in photosynthesis by plants provided with carbon dioxide labelled with ^{14}C. Here again determination of radioactivity through the plant showed a greater rate of translocation of sugar from the leaves of plants supplied with boron than was so with boron-deficient plants.

The work of McIlrath and Palser (1956) also indicated that in tomato, and in turnip as well, translocation was reduced in boron-deficient plants as compared with that in normal plants, although with cotton no such consequence of boron deficiency was obvious. Plants were grown in sand cultures either without added boron or containing boron in a concentration of 5 p.p.m. There was observed a lower concentration of sugar and a higher concentration of starch in the boron-deficient turnip and cotton plants but in tomato there was no difference in the sugar concentration of the boron-deficient and boron-sufficient plants, while the starch content of the former was lower than that of the latter. But the roots of boron-deficient tomato and turnip plants had a smaller carbohydrate content than those of normal plants, a difference attributed to reduced translocation in the former resulting from necrosis of the phloem, and it was significant that in cotton, where necrosis of the phloem was not observed as a result of boron deficiency (see p. 86) no difference was observed in the carbohydrate content of the roots of normal and boron-deficient plants. The conclusion drawn was therefore that the reduced rate of translocation resulting from boron deficiency was a consequence of the necrosis of the phloem, that is, of the actual channel of translocation.

The work of Wadleigh and Shive (1939) on cotton seedlings is also important in this connexion. They examined the effects of boron deficiency in the seedlings by means of microchemical tests. For this purpose seedlings were grown in water cultures with and without a supply of boron. The first internal symptom observed was the increased acidity of a few cells scattered through the pith and cortex, the pH of these cells being from 3·8 to 4·4 as compared with the normal value of 5·8 to 6·4. As boron deficiency increased so did the number of these abnormally acid cells, which then also appeared in the pericycle and the older xylem parenchyma. When the majority of the cells of the pith had become very acid their cell walls began to break down, at the same time developing a deep brown colour. Next, some of the cells of the phloem and younger xylem parenchyma developed high acidity and ultimately a breakdown of cells occurred in these regions also.

While these changes were proceeding accumulation of sugars was observed in the boron-deficient plants, while starch was abnormally abundant in the endodermis. In the cells of the stem apices the nitrate-nitrogen content was much lower in boron-deficient than in normal plants. This was attributed to failure of nitrate absorption owing to death of the root apices. There was a very marked accumulation of ammonia nitrogen in the cells which developed high acidity. Wadleigh and Shive concluded from the fact that both sugars and ammonia nitrogen accumulate in boron-deficient plants that boron deficiency brings about a decreased rate of oxidation of sugars and of amination of carbohydrate derivatives so that protein substances necessary for maintenance of protoplasm are not formed. Microchemical tests for proteins supported this view, for Millon's reagent, the xanthoproteic test and the biuret test all gave immediate results with the abnormally acid cells of boron-deficient plants, whereas in normal plants pre-treatment with ether and alcohol was necessary to denature the proteins before these tests gave a positive result, thus indicating the degeneration of the proteins of the cells with high acidity and ammonia nitrogen content. The disturbance in the carbohydrate and nitrogen metabolism of the boron-deficient plants may be attributed to a disturbance of the normal oxidation-reduction relations of the cells, and it has been pointed out by Johnston and Dore (1929) and Eaton (1935) that boron has considerable affinity for the hydroxy groups of poly-hydric alcohols.

Steinberg (1955) found that boron deficiency in tobacco resulted in a marked increase in the content of nicotine and this appeared to be related to excessive branching of the root system which is also a symptom of boron deficiency in this plant. This was held to support the view that nicotine formation is associated with root and shoot apices. It would also suggest that boron is connected in some way with nitrogen metabolism.

Heggeness (1942) has suggested that boron may play a considerable part in protecting flax from attack by the rust *Melampsora lini*. Borax was applied to the soil at the rate of 60 lb per acre, 2 weeks after germination of the flax, which was sown on 9 May 1941. Some 10 weeks later plants on control plots which did not receive boron showed 100 per cent infection, while plants on the boron-treated plots showed very little infection in the field. When leaves from boron-treated plants were cleared with 80 per cent alcohol many points of infection could indeed be seen, but they mostly failed to develop. With a sowing later in the year, made on 16 June, the boron-treated plants were not so free from rust, but were yet very much freer from rust than the controls.

CHAPTER VII

TRACE ELEMENTS IN PLANTS
IN RELATION TO SOME DISEASES OF
GRAZING ANIMALS

IN some instances deficiency or excess of an element in the plants eaten by herbivorous animals may cause grave disorders in the animals concerned. Deficiency of any essential element will, of course, induce pathological effects, while practically any trace element may produce poisoning if administered or presented to the animal in too great excess. But a few disorders traceable to mineral deficiency in pasture plants, or to excess of certain elements in the herbage eaten by livestock, call for particular mention in an account of trace elements in plants. A deficiency disease is caused by shortage of cobalt in herbage; diseases due to excess of a mineral constituent in the herbage are those in which the pathological effects result from a high proportion of selenium or molybdenum in the pasture plants. For a general account of mineral deficiencies and excesses in pastures in relation to animal nutrition reference may be made to a monograph on the subject by Russell (1944).

1. DISEASES DUE TO TRACE-ELEMENT EXCESS

Alkali disease (selenium poisoning). For many years a disease affecting horses, cattle, pigs and poultry has been known to occur in certain areas of the great plains of the Middle Western United States. The symptoms of the disease include loss of hair from the mane and tail of horses, from the switch of cattle and from the body of pigs, and a marked change in the growth of the horn of the hoof in all these animals which, when severe, results in the sloughing of the hoof. General loss of condition involving emaciation and listlessness may follow, and in the worst cases the animals may die. There may be lesions in various internal organs, including the heart, liver, spleen and kidneys. With affected poultry eggs do not hatch. Further, the symptoms can develop in animals in non-affected areas when fed with hay or grain produced in the affected areas. Because of the popular opinion that these effects were due to alkali in the water drunk by the animals, the disease was generally known as 'alkali disease', but nearly fifty years ago careful work by Larsen, White and Bailey (1912, 1913) showed that this conclusion was false and the cause of the disease must be sought elsewhere. That this was to be found in poisoning by selenium appears to have been first suggested in 1931 (see Byers, 1934). First investigations showed that selenium was present in the soils of the affected areas and that wheat grain from such areas might contain from 10 to 12 p.p.m. of selenium (Robinson, 1933; see also Beath *et al.* 1934, 1935; Byers and Knight, 1935). Nelson, Hurd-Karrer and Robinson (1933) found that wheat appeared to grow normally on soil to which 1 p.p.m. of selenium had been added in the form of sodium selenate, but the grain from this wheat was toxic to white rats fed with it, retardation of growth, and finally death, resulting. Schoening (1936) produced the typical symptoms of alkali disease in hogs by feeding them with maize grown in the affected area. Two lots of maize were used, one containing 5 p.p.m. the other 10 p.p.m. of selenium.

Further experiments on dosing animals with selenium have left no doubt that this element is responsible for producing the symptoms of alkali disease, and that this is, in fact, selenium poisoning. The animals used included rats (Franke and Moxon, 1936), rabbits (Smith, Stohlman and Lillie, 1937), swine (Miller and Schoening, 1938; Miller and Williams, 1940a), cattle, horses and mules (Miller and Williams, 1940a). As an example of the type of experiment employed may be cited one by Miller and Schoening on swine. In this eight pigs about 4 months old were separated into four pairs. All received a sufficient ration of grain, but the four groups received different amounts of selenium, the proportions of this, as sodium selenite in the ration, being respectively 392, 196, 49 and 24·5 p.p.m. Four of the animals developed typical symptoms of alkali disease, namely, loss of hair and disturbance of the growth of the hoof, and all died in from 10 to 99 days. Post-mortem examination revealed lesions of internal organs, particularly of the liver, while kidneys, heart and spleen were also affected in some cases. Two control animals fed with similar grain without added selenium developed normally.

The minimum lethal dose of selenium for rats, rabbits, horses and mules appears to be of the order of 1·5 or 2 mg per lb of body weight, for cattle 4·5–5 mg per lb of body weight, and for pigs between 6 and 8 mg per lb of body weight.

By the continued administration of small doses of selenium to horses Miller and Williams (1940b) produced symptoms similar to, though not so striking as, those observed in chronic alkali disease under natural conditions. These symptoms included listlessness, loosening of hair in the mane and tail, softening of the horny wall of the hoof, and lesions in the liver, heart, kidneys and spleen.

It being thus clear that 'alkali disease' results from poisoning by grain or forage containing selenium, it is important to know whether there are differences in the selenium-absorbing capacity of different plants. There is no doubt that this is very much the case. Byers (1935) found notable differences in the selenium content of plants occurring naturally on seleniferous soils, while Hurd-Karrer (1935) also found marked differences in the selenium content of crop plants grown on artificially selenized soils. Thus Miller and Byers (1937) recorded a selenium content of 1110 p.p.m. in *Astragalus bisulcatus* and one of only 45 p.p.m. in wheat grown in the same area. In another area they found 1250 p.p.m. of selenium in *A. bisulcatus*, but only 3 p.p.m. in *A. missouriensis*. Byers had earlier found as much as 4300 p.p.m. of selenium in *A. bisulcatus*, while relatively enormous quantities of selenium have also been found in other species of *Astragalus*, as, for example, 5560 p.p.m. in *A. ramosus* and 1750 p.p.m. in *A. pectinatus*. On the other hand, *A. missouriensis*, *A. mollissimus* and *A. drummondii* absorb only little selenium. Prairie grasses in general have a very low capacity for absorbing selenium; thus *Andropogon scoparius* (little blue-stem) was found to contain only 0·8 p.p.m. of selenium.

Miller and Byers (1937) distinguished three classes of plants in regard to their capacity for absorbing selenium: (1) highly seleniferous plants which absorb selenium readily and which are either absent from or rare in neighbouring non-seleniferous areas. Plants of this group include *Astragalus bisulcatus*, *A. racemosus*,

A. pectinatus, A. carolineanus, Stanleya pinnata, S. bipinnata, Applopappus fremonti, Xylorrhiza parryi; (2) plants capable of absorbing selenium, even in considerable amount, without severe injury, but which are widely distributed on both seleniferous and non-seleniferous soils. This group includes *Aster ericoides* (white wreath aster), *A. fendleri* (blue aster), *Gutierrezia sarothrae* (turpentine weed), *Helianthus annuus* (sunflower), *Agropyron smithii* (western wheat-grass) and the common cereals, wheat, rye, barley and maize; (3) plants with a low tolerance for selenium, which make poor growth at best on seleniferous soils and which absorb only small amounts of selenium. Plants in this group include *Bouteloua gracilis* (buffalo grass) and *B. curtipendula* (grama grass).

For the control of selenium poisoning the relation of selenium absorption to sulphur absorption by plants may be of great significance. Hurd-Karrer (1934, 1935) showed that increasing the sulphur content of the soil brought about a reduction in the quantity of selenium absorbed by plants. Similarly, with plants grown in water culture selenium uptake was reduced by increasing the concentration of sulphate in the nutrient solution. Hurd-Karrer and Kennedy (1936) found that grain from wheat grown on soil containing 2 p.p.m. of selenium was toxic to white rats, whereas grain from wheat grown on similar selenized soil treated with flowers of sulphur or gypsum was not toxic, the selenium content of the grain being reduced by such treatment from about 12 p.p.m. to about 4 p.p.m.

In 1937 Hurd-Karrer reported the results of experiments on the uptake of sulphur and selenium by some fifteen different crop plants grown in the greenhouse on soil artificially selenized by the addition of 4 p.p.m. of selenium as sodium selenate. The plants were known to differ widely with regard to their capacity for absorbing sulphur. The shoots were cut when 1, 2 and 3 months old and the sulphur and selenium determined. The results obtained after the first month are shown in Table 33. Inspection of this shows that although the species used exhibit a wide range in sulphur and selenium uptake, the ratio of sulphur to selenium absorbed is roughly the same in all.

Hurd-Karrer also found that the sulphur/selenium ratio was practically the same for stems and leaves and also that the ratio

TABLE 33. *Content of sulphur and selenium of a number of crop plants one month old.* (From Hurd-Karrer)

Species	Sulphur (per cent of dry weight)	Selenium (per cent of dry weight)
Brassica oleracea capitata (cabbage)	3·37	0·0793
B. oleracea botrytis (cauliflower)	2·88	0·0710
B. nigra (black mustard)	2·78	0·0744
B. napus (rape)	2·39	0·0780
B. oleracea acephala (kale)	2·24	0·0615
Linum usitatissimum (flax)	1·28	0·0474
Helianthus annuus (sunflower)	1·32	0·0310
Trifolium pratense (red clover)	1·21	0·0315
Vicia villosa (vetch)	0·92	0·0280
Lactuca sativa (lettuce)	0·99	0·0195
Triticum sativum (wheat)	0·87	0·0240
Secale cereale (rye)	0·80	0·0280
Glycine max (Soya bean)	0·59	0·0130
Zea mais (maize)	0·42	0·0140
Sorghum vulgare	0·48	0·0130

depended on the relative amounts of sulphur and selenium in the soil. If the former remained the same while the selenium content was varied, the sulphur/selenium ratio in the plant rose with a fall in the selenium content of the soil. Thus, with a sulphur content of the soil of 0·06 per cent, the average ratio of sulphur to selenium in the plants was found to be 114 when the selenium in the soil was 3 p.p.m., but when this latter was increased to 5 p.p.m. the sulphur/selenium ratio in the plants was only 22.

These results are readily understandable in view of the chemical similarity of sulphur and selenium. They indicate very definitely a way of reducing the selenium content of plants growing on seleniferous soils.

For a good detailed account of the earlier work on selenium poisoning reference may be made to a review by Trelease and Martin (1936).

The scouring of cattle on the teart lands of Somerset. On certain soils mostly derived from the Lower Lias, there are pastures which bring about the complaint of cattle known as scouring. Cattle, on being turned into these pastures, develop the symptoms of scouring within a few days. The symptoms are described by Ferguson, Lewis and Watson (1943) as follows: 'The dung becomes extremely loose and watery, yellowish green in colour, bubbly, and has a foul smell. The animals become filthy, their coats stare and they lose condition rapidly. Red Devon cattle turn a dirty yellow and black beasts go rusty in colour. Milk yields drop considerably.' Sheep are not affected to the same extent, but the dung is very soft and the fleeces sometimes become stained.

The principal area in which teart soils occur is in central Somerset, but smaller areas occur in Warwickshire and Gloucestershire. It is to be noted that an area in Glamorgan in which the soil is also derived from the Lower Lias does not exhibit teartness.

The effect of teart land on cattle has been known, according to Ferguson, Lewis and Watson, for over a hundred years, and has been attributed to a number of factors, including bacteria, parasites, particular species in the herbage, water supply, poor drainage, soil texture, a high proportion of non-protein nitrogen in the herbage, and some particular chemical constituent present in the herbage (Muir, 1936). Bacteria cannot be the primary cause of scouring because the symptoms cease directly the cattle are removed from teart pastures, while as regards parasites there is no abnormally high number of these in affected animals. Nor do the water supply, drainage and soil texture of teart pastures show constant and characteristic differences from non-teart areas. Ferguson, Lewis and Watson determined the nitrogen fractions in herbage from teart and non-teart areas but found no difference in the non-protein nitrogen of the two.

These workers were thus led to investigate the last of the suggestions listed above and made a spectrographic examination of a number of samples of herbage from different sources. As a result they found one constant difference between the samples from teart and from non-teart pastures, namely, in the content of molybdenum, which was considerably higher in the case of the teart herbage. Thus the mean molybdenum content of the herbage from twelve teart localities was 33 p.p.m. of dry matter in 1937 and 38 p.p.m. of dry matter in 1938; the corresponding mean values for eleven non-teart localities were 5 and 4 p.p.m. for the respective years. This finding strongly suggested that molybdenum might be the cause of scouring, and this was confirmed by administering doses of ammonium or sodium molybdate to cattle when the symptoms of scouring were produced in a number of cases, although there was considerable variation in the degree to which different individual animals reacted to molybdenum. Scouring was also produced in cattle and sheep turned into non-teart pasture dressed with sodium molybdate.

Hay and frosted herbage do not have the same effect as the fresh material in inducing scouring. This appears to be due to the fact that fresh herbage contains a much higher proportion of the molybdenum in soluble form than does hay or frosted material, and it would thus appear that it is the soluble fraction of the molybdenum that is responsible for the effect on cattle.

It has been mentioned that the teart soils occur on the Lower Lias where this is directly exposed, but that such an area in Glamorgan does not exhibit the properties of teart land. Lewis (1943a) examined a number of Lower Lias soils in their relation to teartness and found that the teart soils contained about 20–100 p.p.m. of molybdenum in the surface horizon and were neutral or alkaline in reaction. The soils of the Glamorgan area, however, were found to contain only about 2–4 p.p.m. of molybdenum, which explains why they are not teart in character.

Very interesting are the results of experiments made by Lewis (1943b) to examine the uptake of molybdenum from teart soil by plants of a number of different pasture species. Ten pasture grasses and two species of clover were used in these experiments, the seeds being sown in pots of teart soil, four pots of each species being used. The herbage was cut when it was about 3–4 in. high and the molybdenum content determined. After a further period of growth it was again cut and the molybdenum content in the new sample determined. This procedure was repeated so that four samples were obtained in all for each species. The results are summarized in Table 34.

These results show very clearly that the clovers and *Holcus lanatus* contain a much higher proportion of molybdenum than the other pasture plants examined. Teart pastures generally contain very little *Holcus*, but often a good deal of clover. Lewis expressed the opinion that it cannot be assumed that the clovers are the only cause of teartness because analyses made of a number of grasses other than *Holcus* growing on teart land show that these can contain a much higher proportion of molybdenum than was found in the plants grown in pots (cf. Table 35).

TABLE 34. *Uptake of molybdenum by pasture grasses and clovers.* (Data from Lewis)

Molybdenum content
(p.p.m. dry matter)

Species	May–June 1939	July–Aug. 1939	Sept. 1939	May–June 1940
Holcus lanatus (Yorkshire fog)	36	83	61	42
Agrostis alba (fiorin)	14	10	10	9
Phleum pratense (timothy)	8	8	4	6
Festuca pratensis (meadow fescue)	15	10	6	9
Dactylis glomerata (cocksfoot)	13	17	15	9
Poa trivialis (rough-stalked meadow grass)	12	18	15	11
P. pratensis (smooth-stalked meadow grass)	8	6	6	8
Cynosurus cristatus (crested dogstail)	10	8	5	9
Lolium perenne (indigenous perennial ryegrass)	11	11	11	11
L. italicum (Italian ryegrass)	12	13	11	10
Trifolium repens (wild white clover)	93	109	69	57
T. pratense (wild red clover)	87	103	62	59

TABLE 35. *Molybdenum content in p.p.m. of dry matter of pasture plants growing on teart land.* (Data from Lewis)

Species	Soil type Ham-bridge	Eve-sham	Haselor
Dactylis glomerata (cocksfoot)	70	42	13
Poa pratensis (smooth-stalked meadow grass)	54	39	18
Lolium perenne (perennial ryegrass)	54	36	11
Cynosurus cristatus (crested dogstail)	60	36	—
Bromus mollis (soft brome)	60	11	18
Agrostis spp. (bent)	54	41	—
Phleum pratense (timothy)	—	30	8
Festuca spp. (fescues)	—	—	20
Arrhenatherum avenaceum (tall oat grass)	—	—	12
Trifolium spp. (clovers)	90	156	28

Subsequently malnutrition of cattle as a result of grazing in pastures of high molybdenum content has been recognized as occurring in many parts of the world including Sweden, Canada, the United States and New Zealand (cf. Dick, 1956). In America and elsewhere it has sometimes been called molybdenosis and in New Zealand peat scours.

Ferguson, Lewis and Watson found that the scouring of cattle and sheep resulting from the absorption of molybdenum can be prevented and cured by dosing the animals with copper sulphate. On very teart land a daily dose of 2 g of this salt for cows and 1 g for young stock was found to be adequate for the purpose, while on mildly teart land a smaller dose would probably be sufficient. That the harmful effects of molybdenum excess can be counteracted by giving the affected animals copper sulphate suggests that molybdenum inhibits either the absorption or utilization of copper. Further evidence of this is suggested by the occurrence in Holland of a disease attributed by Sjollema (1938) to shortage of copper, the symptoms of which are similar to those produced by excess of molybdenum, as they were described as diarrhoea, wasting and, in black cattle, loss of colour in the coat which becomes brown-grey. The copper content of the blood, liver and milk of affected cows and goats was abnormally low, and so was the hair of affected cows. The grass and hay from farms on which the disease occurred had a low content of both copper and manganese, but the quantity of the latter appeared more than adequate for the needs of the animals. The hay was also poor in zinc, but the blood of affected animals contained a normal amount of this element and the liver a higher content than the normal. As regards iron the content of this element in diseased animals was abnormally high. Finally, dosing with copper sulphate brought about the recovery of the affected animals.

While there can be little doubt that molybdenum has a very definite effect on the copper metabolism of ruminants it is not clear what the interaction of these metals is. Davis (1958) suggested that molybdenum tended to suppress the action of the copper-containing enzyme tyrosinase which would affect the colour of the animal's coat. He also thought the mobilization of phosphorus in the bones might be concerned. Further, he thought that molybdenum also affected cobalt metabolism, for the synthesis of vitamin B_{12} was markedly depressed if more than 40 p.p.m. of molybdenum was present in the animal's diet. It was thought that the effect of molybdenum was on the organisms in the rumen which synthesized vitamin B_{12} from cobalt.

Ways to reduce the teartness of pastures were also examined by Lewis. He found that application of acidic nitrogenous fertilizers containing ammonium sulphate reduced the proportion of molybdenum in the herbage. This was largely due to a reduction in the proportion of clover in the herbage, but the actual proportion of molybdenum in the grasses was also reduced. This might have been due to the ammonium sulphate bringing about an increase in the yield of the grasses without increasing the weight of molybdenum taken up. Also it has been noted that teart soils are neutral or alkaline. Rendering the soil more acidic will reduce the availability of the molybdenum and so induce a lowering in molybdenum absorption by the plants.

Another point observed by Lewis was that the molybdenum content of newly sown grasses is low, but increases with age, although clovers have a high molybdenum content from the beginning. A system of ley farming with short leys and with only a small percentage of clover would thus appear to be very suitable for teart land.

2. A TRACE-ELEMENT DEFICIENCY DISEASE OF ANIMALS

Pining, enzootic marasmus, bush sickness or Morton Mains disease. In countries as far apart as Scotland and Australia and New Zealand there has occurred for many years a disease of sheep characterized by the rapid deterioration of the animal. In Scotland, where the disease occurs in many parts of the country, it is known as pine or pining, vinquish and daising, and young cattle are also affected. Here the disease was recorded by Hogg, the Ettrick Shepherd, as long ago as 1831. In recent years pining in sheep has been recognized as occurring in various parts of England, notably in areas of Northumberland and Cumberland, and on Exmoor and Dartmoor in Devonshire. The symptoms as they appear in the island of Tiree, Inner Hebrides, have been described by Greig, Dryerre, Godden, Crichton and Ogg (1933) as 'those of a progressive debility, accompanied by anaemia and emaciation. The onset is frequently insidious. The affected animal is dull, and the coat, in the case of cattle, is rough and staring, the visible mucous membranes, especially the conjunctiva, become pale: physical condition is gradually lost, the eyeball becomes sunken, and there is commonly a serous discharge from the inner canthus. In young animals growth is markedly retarded, and they soon present a stunted, unthrifty appearance. Thereafter the anaemia and emaciation progress to the condition of cachexia, and, finally, as the result of extreme weakness, the animal is unable to rise. In severe cases the gait is stilted and somewhat incoordinated.' The disease usually ends fatally.

The disease of sheep in North Island, New Zealand, known as bush sickness, that in Southland, New Zealand, known as Morton Mains disease, and that in Australia called enzootic marasmus, all appear to be the same as pine in Scotland. Morton Mains disease in a bad year and in a bad locality was described by Wunsh (1937) thus: 'About midsummer a majority of the season's crop of lambs would fall off in condition. Their wool would become "chalky" and harsh, their eyes would water, and they would lose their activity. A large number—30 per cent or more— would gradually lose weight, become more and more helpless and finally die.' (See also Askew and Rigg, 1932.)

A consideration of the facts known about Morton Mains disease in New Zealand suggests that it is a deficiency disease. Thus the disease does not appear on newly broken land, but only after several years of sheep farming on the same land. Also deep ploughing lessens the disease for a few years. Both these facts suggest a deficiency, and certainly not an excess, of some substance as the cause of the disease. That the disease is not due to bacteria or other parasites is indicated by the fact that a sick lamb transferred to healthy country recovers rapidly. Since the climate, physical conditions and type of herbage of healthy and affected farms might show no appreciable differences, there is a strong suggestion that some substance in very small amount, such as a trace element, is involved.

In 1935 it was reported by Underwood and Filmer that enzootic marasmus of sheep in Western Australia could be cured by administering a dose of 0·1–2 mg of cobalt nitrate each day, and 2 years later it was reported that affected cattle could be cured in the same way (Filmer and Underwood, 1937). In the meantime Askew and Dixon (1936) had reported a similar favourable effect of cobalt on sheep suffering from Morton Mains disease in New Zealand. This clearly suggested that deficiency of cobalt might be the cause of the disease, and this view was supported by the results of analyses of some New Zealand soils made by Ramage by means of his spectrographic technique (see p. 25), and which showed that two healthy soils each contained 7 p.p.m. of cobalt while a sick Morton Mains soil contained none of this element. Subsequent treatment of affected lambs in New Zealand with weekly amounts of 7 mg of cobalt brought about a remarkable improvement after a fortnight.

In 1938 Underwood and Harvey showed that both soil and herbage of affected areas in Australia have a lower cobalt content than those of neighbouring healthy areas, but that the cobalt content of the herbage is considerably increased by dressing the soil with a little cobalt acetate. Similar results have been obtained in New Zealand (cf. Kidson and Maunsell, 1939), although it would appear that there is not always a direct relation between cobalt contents of soil and herbage.

Experiments in Scotland on pining have similarly shown the relationship of this disease to a deficiency of cobalt. In south-east Scotland Corner and Smith (1938) have shown that pining could be cured and its reappearance prevented for 6 months by a daily dose of 1 mg of cobalt for 14 days.

An examination of a number of soils in the north of Scotland, an area where pining occurs, made by Stewart, Mitchell and Stewart (1941), showed that the cobalt content of the soils varied from 1 to 300 p.p.m., and that most of the soils on which pining occurs have a cobalt content of less than 5 p.p.m. while some contain as little as 1 p.p.m. The same workers also carried out an experiment on the treatment of affected lambs with cobalt. In this experiment, which was started towards the end of June 1939, two sets of lambs were used, one comprising forty individuals which were treated with cobalt, the other of twenty-five which served as controls, and which at the beginning of the experiment were the best in the flock. Both groups ran together on bad pining land in which the cobalt content of the soil was only 1–2 p.p.m. The lambs of the experimental group were each given 10 mg of cobalt as cobalt chloride each week for 10 weeks; the control animals received no cobalt. Apart from one lamb which died from an unknown cause 3 days after the beginning of the experiment and which was immediately replaced by one from the control group, all the treated animals improved in condition and growth and were stated by the farmer to be the best set of lambs he had seen on the farm where the experiment was carried out. The control animals, on the other hand, rapidly lost condition, four died from pining, and by 4 September many others were in the last stage of the disease.

Stewart, Mitchell and Stewart also examined on a set of plots the effect on the cobalt content of soil and herbage which resulted from the application of various quantities of cobalt chloride to the soil, the amounts ranging from $\frac{1}{2}$ to 80 lb of cobalt chloride (CoCl$_2$.6H$_2$O) per acre, superphosphate to the extent of $2\frac{1}{4}$ cwt per acre being used as filling material for the small quantity of cobalt used. Ten weeks after the application of the dressings the total cobalt content of the soil and the amount soluble in dilute acetic acid were determined on each plot as well as the cobalt content of the herbage. The results, given in Table 36, show how in this comparatively short time the application of cobalt to the soil affected the cobalt content of the herbage.

TABLE 36. *Effect of cobalt manuring on the cobalt content of herbage.* (Data from Stewart, Mitchell and Stewart)

Cobalt applied to soil (p.p.m.)	Cobalt in soil (p.p.m.)		Cobalt in herbage (p.p.m.)
	Total	Readily soluble	
0	2·9	0·177	0·11
0	{ 2·9	0·196	0·12
	{ 2·8	0·200	0·20
0·06	—	—	{ 0·23
			{ 0·35
0·125	{ 2·9	0·215	0·34
	{ 2·9	0·230	0·41
0·25	{ 3·25	0·255	0·51
	{ 3·2	0·236	0·83
1·25	{ 4·0	0·479	1·63
	{ 4·2		2·28
5·00	5·8	1·75	3·94
10·00	9·6	1·96	8·17

The favourable effect on lambs produced by such manuring was shown by another experiment carried out by Stewart, Mitchell and Stewart in which a field with a low cobalt content of the soil (about 1·8 p.p.m.) was divided into two halves, one of which was manured with 2 lb cobalt chloride + 3 cwt superphosphate per acre, the other with 3 cwt superphosphate only. Four to five weeks after this manuring fifteen lambs were allowed to run on each half of the field, the two groups being of equal value at the start. After 6 weeks the difference in the appearance of the lambs of the two groups was striking, all those on the half of the field with added cobalt being in fine condition, whereas eight of the fifteen lambs on the other half of the field showed severe pining while the remaining seven animals were in poor condition.

While the effectiveness of cobalt manuring in increasing the cobalt content of the herbage of 'pining' lands can be regarded as established, work by Mitchell, Scott, Stewart and Stewart (1941) indicates that such manuring requires care. They found by spectrographic analysis that the herbage growing on cobalt-treated soil could take up an abnormally large amount of molybdenum. Where the herbage without cobalt treatment already had a fairly high molybdenum content, this latter might be so increased by cobalt manuring that it approached the molybdenum content of the herbage of teart land, so that danger of scouring could ensue. Analyses of the herbage from two soils, one with a low, the other with a high molybdenum content, cut 15 months after various degrees of cobalt manuring, are shown in Table 37. Since Lewis has shown that molybdenum uptake is reduced by the use of acidic nitrogenous fertilizers (see p. 188), and since teartness occurs on neutral and alkaline soils, it would seem inadvisable to apply cobalt in the form of a cobalt-rich lime or in conjunction with fertilizers having an alkaline reaction.

TABLE 37. *Cobalt and molybdenum contents of herbage from soils treated with cobalt chloride.* (Data from Mitchell, Scott, Stewart and Stewart)

$CoCl_2.6H_2O$ added (lb per acre)	Herbage from soil A		Herbage from soil B	
	Co (p.p.m.)	Mo (p.p.m.)	Co (p.p.m.)	Mo (p.p.m.)
0	0·08	1·7	0·07	6·3
2	0·22	2·2	0·20	9·2
10	0·63	2·4	0·89	10·0
80	3·20	7·5	2·75	14·2

A disease of cattle known as 'salt sick' has been known for many years in Florida. This disease was considered by Becker, Neal and Shealy (1931) to be a nutritional anaemia resulting from an insufficiency of copper and iron in the diet. Bryan and Becker (1935) reported that the disease occurred on certain sandy soils, the surface layers of which possessed roughly only one-tenth of the iron content and one-half of the copper content of healthy soils of Florida. They found that cattle became salt sick on soils containing 0·036 per cent of iron and 3·85 p.p.m. of copper, and remained healthy on soils containing 0·42 per cent of iron and 8 p.p.m. of copper. It was suggested by Aston (1931) and Greig, Dryerre, Godden, Crichton and Ogg (1933) that the disease was identical with pining. And more recently Thacker and Beeson (1958) stated that salt sick, burton ail, pining, vinquish, bush sickness, wasting disease, coast disease and nakuritis have all proved to be local descriptive names for the disease in cattle and sheep due to inadequate intake of cobalt which results when the forage contains less than 0·07 p.p.m. of that element. A disease of cattle known as licking sickness, described by Sjollema (1933) as occurring on farms in Holland where reclamation disease due to copper deficiency occurs, was attributed by him to copper shortage, but the general description of the symptoms of the disease bears a strong resemblance to that of pining and is unlike that of the symptoms of the scouring disease of cattle attributed to molybdenum excess which, as we have seen, probably affects the absorption or utilization of copper. Spectrographic analysis of grasses from salt-sick and healthy areas made by Rusoff, Rogers and Gaddum (1937) lent no support to the view that salt sick resulted from copper shortage for they found the same content of copper in the grasses of salt-sick areas as in those of healthy areas.

The importance of an adequate supply of cobalt depends on the presence of this element in vitamin B_{12} and it is without doubt the inadequate amount of this vitamin which brings about the anaemic condition of the animals suffering from pining.

CHAPTER VIII

CONCLUDING REMARKS

I T will be clear from the preceding review of the present position of our knowledge of trace elements in plants that these elements present the biologist with two sets of problems, the one pathological, the other physiological. The pathologist is concerned with the abnormal conditions resulting in plants from deficiency or excess of the various trace elements, and of the means by which the deficiency or excess can be removed. The problems of the physiologist are more subtle, for it is his business to discover the functions in the life of the organism of these various elements which are present in only minute amounts. The two sets of problems are of course interdependent, for the work of the pathologist in discovering the effects of deficiency greatly aids the physiologist, while knowledge of the part played by the trace elements in the life of the organism must necessarily help the pathologist in the diagnosis and treatment of deficiency diseases. But on the whole, knowledge of the pathology of trace elements is much more advanced than our knowledge of their physiological functions. A considerable number of well-recognized and defined plant diseases are now correctly attributed to various trace-element deficiencies, and means of controlling these diseases have been determined with considerable precision. Future work on the pathological side may extend the number of such known deficiency diseases. In addition to such work it would appear that two lines of research in the field of pathology would well repay attention. The first of these is the development of rapid means of early diagnosis. The usual method of observing deficiency symptoms, employed with outstanding success by Wallace, is naturally only applicable after visible external symptoms have developed. The injection methods elaborated by Roach appear to be particularly useful for trees, but although certainly usable for annual crops, and affording a means of early diagnosis, require some degree of technical skill for their employment with small herbaceous plants.

The second field of research which should afford valuable results in the control of deficiency diseases is the investigation of the relationship of soil and climatic factors to the incidence of these diseases. It has been pointed out in the course of the preceding pages how the amount of any trace element in the soil directly available for absorption by plants is generally only a fraction of the total amount present and that the degree of availability may be affected by various factors, of which the most obvious is the acidity of the medium. With decreasing acidity (increasing pH) manganese, zinc and boron all become less available, whereas molybdenum is affected in the opposite way, more going into solution as the soil reaction becomes more alkaline. Hence liming the soil, by bringing about an increase in the pH, may significantly affect the amount of every one of the micronutrients and a treatment which is beneficial to plants in respect of one micro-nutrient may be detrimental in respect of another, as, for example, if manganese were present in excessive amount and boron in a very low concentration. Other factors have also been shown to influence availability, a notable one being the humus content of the soil. The actual relations between pH, humus content and other soil conditions on the one hand, and the concentration of available trace elements on the other, are still very far from being completely understood. Again, it is well established that the severity of deficiency diseases may vary in different years, a variation which must be attributed to the differences in climate experienced in different seasons; the symptoms of boron deficiency, for example, are generally much more severe in periods of drought than in wetter seasons. It is to be presumed that the different water contents of the soil in the different seasons must affect the availability of the trace elements concerned.

Nor must the micro-organisms of the soil be forgotten. These may themselves play a part in determining the availability of various elements by bringing about oxidations and reductions of compounds of the different trace elements and so affecting their solubility, or they may play a more direct part in the deficiency diseases themselves, as suggested by the work of Gerretsen and Ark referred to in an earlier chapter. There is here a wide field of investigation which at present has scarcely been entered.

In the physiological field the outstanding fact is, of course, that the trace elements as their designation implies are needed in very minute amount. Because of the small quantities of them involved in the organism it was natural to suppose that they acted as catalysts, and one of the outstanding advances made in our knowledge of the functions of trace elements during the last twenty years has been the demonstration that they enter into the composition of a considerable number of enzymes—manganese, zinc, copper and molybdenum all being concerned in this way. In attempting to determine the part played by the individual trace elements in the life of the plant, while it is reasonable to suppose that they perform the same function in enzyme actions in the living cell as they do in non-living systems in which their role has been demonstrated, the extreme complexity of the mechanism of the living cell has to be borne in mind. The conception of a balance between various major nutrients as determining plant growth is now an old one, and there is abundant evidence of subtle interrelationships not only between individual trace elements and major nutrient elements, but between trace elements themselves. The relation between manganese and iron has been dealt with in some detail in these pages but many other instances of the interaction of trace elements with one another are on record. For example, Millikan (1947) found that flax grown in water culture exhibited symptoms of iron deficiency if the nutrient solution contained manganese, zinc, copper, cobalt or nickel in equivalent concentration but did not contain molybdenum, but that if molybdenum was present the severity of the symptoms was progressively lessened with increase in the molybdenum concentration from 2 to 25 p.p.m. Hewitt (1953) grew plants of marrowstem kale, sugar beet, potato, tomato and oats in sand cultures provided with a number of heavy metals in equivalent concentration in addition to the major nutrients. As well as manganese, zinc, copper and molybdenum a number of presumed non-essential elements were used, namely, chromium, cobalt, nickel, lead, cadmium and vanadium. Effects of individual elements were found to differ with different species. Thus although copper induced no chlorosis in oats and kale, it did in sugar beet and tomato, in which plants the chlorosis resembled that produced by deficiency of iron. This type of chlorosis was also produced by cobalt and cadmium. Nickel induced symptoms in potato and tomato resembling those of manganese deficiency as did zinc with sugar beet. No relationship could be detected between such effects and the oxidation-reduction potentials of the metal ions. Another example of the interaction of trace elements was afforded by the observation of Heintze (1956) that application of manganese to a manganese-deficient soil induced symptoms of copper deficiency in oats, although without the addition of manganese there was no indication of copper deficiency. Further investigation of such interrelationships would be likely ultimately to throw much light on trace-element functions.

Whether the metallic trace elements are concerned in plant processes other than enzyme actions is by no means clear, but the information at present available suggests that the action of boron is of a different kind from that of other micro-nutrients, and as our knowledge stands at present it would seem most likely that boron plays a part in carbohydrate metabolism but not as an essential constituent of an enzyme system.

A mode of attack on the physiological aspects of trace elements lies in growing plants under experimental conditions in which the supply of various mineral nutrients is controlled and determining the resultant effects on growth and the fate in the plant of the mineral elements concerned. For the latter the use of radioactive tracers would give promise of fruitful results, but although a few researches have been reported in which such tracers have been employed, as, for example, by Stout and Meagher (1948) with labelled molybdenum and by Sisler, Duggar and Gauch (1956) in work with boron, relatively little work with micro-nutrients has so far been carried out along such lines.

Our knowledge of the part played by trace elements in the life of the plant is now very much greater than it was when the first edition of this book was in preparation some sixteen years ago and some functions of trace elements in relation to the essential part they play in enzyme actions are firmly established. Yet our knowledge of their action is still very fragmentary and it may seem remarkable that the mode of linkage of their functions with the symptoms of their deficiency and excess is still largely unexplained. The trace elements still provide the plant physiologist with a wide field for exploration.

LIST OF LITERATURE

AGULHON, H. (1910). Emploi du bore comme engrais catalytique. *C.R. Acad. Sci.*, *Paris*, **150**, 288–91.

ALBEN, A. O., COLE, J. R. and LEWIS, R. D. (1932*a*). Chemical treatment of pecan rosette. *Phytopathology*, **22**, 595–601.

ALBEN, A. O., COLE, J. R. and LEWIS, R. D. (1932*b*). New developments in treating pecan rosette with chemicals. *Phytopathology*, **22**, 979–80.

ALEXANDER, T. R. (1942). Anatomical and physiological responses of squash to various levels of boron supply. *Bot. Gaz.* **103**, 475–91.

AMIN, J. V. and JOHAM, H. E. (1958). A molybdenum cycle in the soil. *Soil Sci.* **85**, 156–60.

ANDERSON, A. J. and OERTEL, A. C. (1946). Factors affecting the responses of plants to molybdenum. *Bull. Coun. Sci. Industr. Res. Aust.* no. 198, 25–44.

ANDERSON, A. J. and SPENSER, D. (1950). Molybdenum in nitrogen metabolism of legumes and non-legumes. *Aust. J. Sci. Res.* B, **3**, 414–30.

ANDERSON, A. J. and THOMAS, M. P. (1946). Plant responses to molybdenum as a fertilizer. I. *Bull. Coun. Sci. Industr. Res. Aust.* no. 198, 7–24.

ANDERSON, I. and EVANS, H. J. (1956). Effect of manganese and certain other metal cations on iso-citric dehydrogenase and malic enzyme activities in *Phaseolus vulgaris*. *Plant Physiol.* **31**, 22–8.

ANDERSSEN, F. G. (1932). Chlorosis of deciduous fruit trees due to a copper deficiency. *J. Pomol.* **10**, 130–46.

ARGAWALA, S. C. (1952). Relation of nitrogen supply to the molybdenum requirement of cauliflower grown in sand culture. *Nature, Lond.*, **169**, 1099.

ARK, P. A. (1937). Little-leaf or rosette of fruit trees. VII. Soil microflora and little-leaf or rosette disease. *Proc. Amer. Soc. Hort. Sci.* 1936, **34**, 216–21.

ARNON, D. I. (1937). Ammonium and nitrate nitrogen nutrition of barley at different seasons in relation to hydrogen-ion concentration, manganese, copper and oxygen supply. *Soil Sci.* **44**, 91–121.

ARNON, D. I. (1938). Microelements in culture-solution experiments with higher plants. *Amer. J. Bot.* **25**, 322–5.

ARNON, D. I. (1940). The essential nature of molybdenum for the growth of higher plants. *Chron. Bot.* **6**, 56–7.

ARNON, D. I., ICHIOKA, P. S., WESSEL, G., FUJIWARA, A. and WOLLEY, J. T. (1951). Molybdenum in relation to nitrogen metabolism. I. Assimilation of nitrate nitrogen by Scenedesmus. *Physiol. Plant.* **8**, 538–51.

ARNON, D. I. and STOUT, P. R. (1939*a*). The essentiality of certain elements in minute quantity for plants with special reference to copper. *Plant Physiol.* **14**, 371–5.

ARNON, D. I. and STOUT, P. R. (1939b). Molybdenum as an essential element for higher plants. *Plant Physiol.* 14, 599–602.

ARNON, D. I. and WESSEL, G. (1953). Vanadium as an essential element for green plants. *Nature, Lond.*, 172, 1039–40.

ASKEW, H. O., CHITTENDEN, E. T. and MONK, R. J. (1951). 'Die-back' in raspberries. *J. Hort. Sci.* 26, 268–84.

ASKEW, H. O., CHITTENDEN, E. and THOMSON, R. H. K. (1936). The use of borax in the control of 'internal cork' of apples. *N.Z. J. Sci. Tech.* 18, 365–80.

ASKEW, H. O. and DIXON, J. K. (1936). The importance of cobalt in the treatment of certain stock ailments in the South Island, New Zealand. *N.Z. J. Sci. Tech.* 18, 73–92.

ASKEW, H. O. and RIGG, T. (1932). Bush sickness. Investigations concerning the occurrence and cause of bush sickness in New Zealand. *Bull. N.Z. Dep. Sci. Industr. Res.* 32, 5–62.

ASKEW, H. O. and WILLIAMS, W. R. L. (1939). Brown-spotting of apricots, a boron-deficiency disease. *N.Z. J. Sci. Tech.* A, 21, 103–6.

ASO, K. (1902). On the physiological influence of manganese compounds on plants. *Bull. Coll. Agric., Tokyo*, no. 5, 177–85.

ASTON, B. C. (1929). Cure of iron starvation (bush sickness) in stock. *N.Z. J. Agric.* 38, 232–7.

ASTON, B. C. (1931). Recent work on iron starvation in other countries. *N.Z. J. Agric.* 43, 270–2.

ATACK, F. W. (1915). A new reagent for the detection and colorimetric estimation of aluminium. *J. Soc. Chem. Ind., Lond.*, 34, 936–7.

BAILEY, L. F. and McHARGUE, J. S. (1944). Effect of boron, copper, manganese and zinc on the enzyme activity of tomato and alfalfa plants grown in the greenhouse. *Plant Physiol.* 19, 105–16.

BAKER, J. E., GAUCH, H. G. and DUGGAR, W. M. (1956). Effects of boron on the water relations of higher plants. *Plant Physiol.* 31, 89–94.

BARNETTE, R. M. and WARNER, J. D. (1935). A response of chlorotic corn plants to the application of zinc sulfate to the soil. *Soil Sci.* 39, 145–56.

BARSHAD, I. (1948). Molybdenum content of pasture plants in relation to toxicity to cattle. *Soil Sci.* 66, 187–95.

BARSHAD, I. (1951a). Factors affecting the molybdenum content of pasture plants. I. Nature of soil molybdenum, growth of plants and soil pH. *Soil Sci.* 71, 297–313.

BARSHAD, I. (1951b). Factors affecting the molybdenum content of pasture plants. II. Effect of soluble phosphates, available nitrogen and soluble sulfates. *Soil Sci.* 71, 387–98.

BAUMANN, A. (1885). Das Verhalten von Zinksalzen gegen Pflanzen und im Boden. *Landw. Versuchs-Stat.* 31, 1–53.

BEATH, O. A., DRAIZE, J. H., EPPSON, H. F., GILBERT, C. S. and McCREARY, O. C. (1934). Certain poisonous plants of Wyoming activated by selenium and their association with respect to soil types. *J. Amer. Pharm. Ass.* 23, 94–7.

BEATH, O. A., EPPSON, H. F. and GILBERT, C. S. (1935). Selenium and other toxic minerals in soils and vegetation. *Bull. Wyo. Agric. Exp. Sta.* no. 206, 56 pp.

BECKER, R. B., NEAL, W. M. and SHEALY, A. L. (1931). I. Salt sick: its cause and prevention. II. Mineral supplements for cattle. *Bull. Fa Agric. Exp. Sta.* no. 231, 22 pp.

BEESON, K. C., GRAY, L. and ADAMS, M. B. (1947). The absorption of mineral elements by forage plants. I. The phosphorus, cobalt, manganese and copper content of some common grasses. *J. Amer. Soc. Agron.* **39**, 356–62.

BEESON, K. C., GRAY, L. and HAMNER, K. C. (1948). The absorption of mineral elements by forage plants. II. The effect of fertilizer elements and liming materials on the content of mineral nutrients in soybean leaves. *J. Amer. Soc. Agron.* **40**, 553–62.

BEEVERS, H. and JAMES, W. O. (1948). The behaviour of secondary and tertiary amines in the presence of catechol and Belladonna catechol oxidase. *Biochem. J.* **43**, 636–9.

BERENBLUM, I. and CHAIN, E. (1938). An improved method for the colorimetric determination of phosphate. *Biochem. J.* **32**, 295–8.

BERGER, K. C. and GERLOFF, G. C. (1947). Manganese toxicity of potatoes in relation to strong soil acidity. *Proc. Soil Sci. Soc. Amer.* **12**, 310–14.

BERGER, K. C. and TRUOG, E. (1939). Boron determination in soils and plants using the quinalizarin reaction. *Industr. Engng Chem.* (Anal. ed.), **11**, 540–5.

BERGER, K. C. and TRUOG, E. (1944). Boron tests and determination for soils and plants. *Soil Sci.* **57**, 25–36.

BERTRAND, D. (1940). Sur la diffusion du molybdène dans la terre arable et dans l'eau. *C.R. Acad. Sci., Paris,* **211**, 406–8.

BERTRAND, D. (1941). Importance de l'oligoélément vanadium pour l'*Aspergillus niger*. *C.R. Acad. Sci., Paris,* **213**, 254–7.

BERTRAND, G. (1897). Sur l'intervention du manganèse dans les oxidations provoquées par la laccase. *C.R. Acad. Sci., Paris,* **124**, 1032–5, 1355–8.

BERTRAND, G. (1905). Sur l'emploi favorable du manganèse comme engrais. *C.R. Acad. Sci., Paris,* **141**, 1255–7.

BERTRAND, G. (1912a). Sur l'extraordinaire sensibilité de l'*Aspergillus niger* vis-à-vis du manganèse. *Bull. Soc. chim. Fr.* IV, **11**, 494–8.

BERTRAND, G. (1912b). Sur le rôle capital du manganèse dans la production des conidies de l'*Aspergillus niger*. *Bull. sci. pharmacol.* **19**, 321–4.

BERTRAND, G. (1912c). Sur le rôle capital du manganèse dans la production des conidies de l'*Aspergillus niger*. *C.R. Acad. Sci., Paris,* **154**, 381–3.

BERTRAND, G. (1940). Importance du molybdène comme oligoélément par les légumineuses. *C.R. Acad. Sci., Paris,* **211**, 512–14.

BERTRAND, G. and JAVILLIER, M. (1911a). Influence combinée du zinc et du manganèse sur le développement de l'*Aspergillus niger*. *C.R. Acad. Sci., Paris,* **152**, 900–3.

BERTRAND, G. and JAVILLIER, M. (1911b). Influence du manganèse sur le développement de l'*Aspergillus niger*. *C.R. Acad. Sci., Paris*, **152**, 225–8.

BERTRAND, G. and JAVILLIER, M. (1911c). Influence du zinc et du manganèse sur la composition minérale de l'*Aspergillus niger*. *C.R. Acad. Sci., Paris*, **152**, 1337–40.

BERTRAND, G. and JAVILLIER, M. (1912a). Action combinée du manganèse et du zinc sur le développement et la composition minérale de l'*Aspergillus niger*. *Ann. Inst. Pasteur*, **26**, 241–6.

BERTRAND, G. and JAVILLIER, M. (1912b). Action du manganèse sur le développement de l'*Aspergillus niger*. *Ann. Inst. Pasteur*, **26**, 241–6.

BERTRAND, G. and JAVILLIER, M. (1912c). Action du manganèse sur le développement de l'*Aspergillus niger*. *Bull. Soc. chim. Fr.* IV, **11**, 212–21.

BINGHAM, F. T., MARTIN, J. P. and CHASTAIN, J. A. (1958). Effects of phosphorus fertilization of California soils on minor element nutrition of *Citrus*. *Soil Sci.* **86**, 24–31.

BISHOP, W. B. S. (1928). The distribution of manganese in plants, and its importance in plant metabolism. *Aust. J. Exp. Biol. Med. Sci.* **5**, 125–41.

BLANK, L. M. (1941). Response of *Phymatotrichum omnivorum* to certain trace elements. *J. Agric. Res.* **62**, 129–59.

BOBKO, E. V. and BELVOUSSOV, M. A. (1933). Importance du bore pour la betterave à sucre. *Ann. Agron.* N.S. **3**, 493–504.

BOBKO, E. V. and SAVVINA, A. G. (1940). Role of molybdenum in plant-development. *C.R. Acad. Sci. U.R.S.S.* **29**, 507–9.

BOKEN, E. (1955). On the effect of ferrous sulphate on the available manganese in the soil. *Plant and Soil*, **6**, 97–112.

BOKEN, E. (1956). On the effect of ferrous sulphate on the available manganese in the soil and the uptake of manganese by the plant. II. *Plant and Soil*, **7**, 237–52.

BOKEN, E. (1957). The effect of ferrous sulphate on the yield and manganese uptake of oats on sandy soil fertilized with pyrolucite. *Plant and Soil*, **8**, 160–9.

BOLLE-JONES, E. W. (1955). The effect of varied nutrient levels on the concentration and distribution of manganese within the potato plant. *Plant and Soil*, **6**, 45–60.

BOLLE-JONES, E. W. (1956). Molybdenum status of laminae as determined by bioassay and chemical methods. *Plant and Soil*, **7**, 130–4.

BORTELS, H. (1927). Über die Bedeutung von Eisen, Zink, und Kupfer für Mikroorganismen. *Biochem. Z.* **182**, 301–58.

BORTELS, H. (1930). Molybdän als Katalysator bei den biologischen Stickstoffbindung. *Arch. Mikrobiol.* **1**, 333–42.

BORTELS, H. (1937). Über die Wirkung von Molybdän- und Vanadium-Düngungen auf Leguminosen. *Arch. Mikrobiol.* **8**, 13–26.

BORTELS, H. (1939). Über die Wirkung von Agar sowie Eisen, Molybdän, Mangan und anderen Spurenelementen in stickstofffreier Nährlösung auf Azotobakter. *Z. Bakt.* II, **100**, 373–93.

BORTELS, H. (1940). Über die Bedeutung des Molybdäns für stickstoffbindende Nostocaceen. *Arch. Mikrobiol.* 11, 155–86.

BOULD, C., NICHOLAS, D. J. D., POTTER, J. M. S., TOLHURST, J. A. H. and WALLACE, T. (1949). Zinc and copper deficiencies of fruit trees. *Ann. Rep. Agric. Hort. Res. Sta., Long Ashton*, pp. 45–9.

BRANDENBURG, E. (1931). Die Herz- und Trockenfäule der Rüben als Bormangelerscheinung. *Phytopath. Z.* 3, 499–517.

BRANDENBURG, E. (1932). Die Herz- und Trockenfäule der Rüben—Ursache und Bekämpfung. *Angew. Bot.* 14, 194–228.

BRANDENBURG, E. (1933). Onderzoekingen over ontginningsziekte. II. *Tijdschr. Plziekt.* 39, 189–92.

BRANDENBURG, E. (1934). Über die Bedeutung des Kupfers für die Entwicklung einiger Pflanzen im Vergleich zu Bor und Mangan und über Kupfermangelerscheinungen. *Angew. Bot.* 16, 505–9.

BRENCHLEY, W. E. (1914). *Inorganic Plant Poisons and Stimulants.* Cambridge.

BRENCHLEY, W. E. and THORNTON, H. G. (1925). The relation between the development, structure and functioning of the nodules on *Vicia faba*, as influenced by the presence or absence of boron in the nutrient medium. *Proc. Roy. Soc.* B, 98, 373–98.

BRENCHLEY, W. E. and WARINGTON, K. (1927). The role of boron in the growth of plants. *Ann. Bot., Lond.*, 41, 167–87.

BROWN, J. C. and HOLMES, R. S. (1955). Iron, the limiting element in a chlorosis. Part I. Availability and utilization of iron dependent upon nutrition and plant species. *Plant Physiol.* 30, 451–7.

BROWN, J. C., HOLMES, R. S. and SPECHT, A. W. (1955). Iron, the limiting element in a chlorosis. Part II. Copper-phosphorus induced chlorosis dependent upon plant species and varieties. *Plant Physiol.* 30, 457–62.

BROYER, T. C., CARLTON, A. B., JOHNSON, C. M. and STOUT, P. R. (1954). Chlorine—a micronutrient element for higher plants. *Plant Physiol.* 29, 526–32.

BRYAN, O. C. and BECKER, R. B. (1935). The mineral content of soil types as related to 'salt sick' of cattle. *J. Amer. Soc. Agron.* 27, 120–7.

BURRELL, A. B. (1937). Boron treatment for a physiogenic apple disease. *Proc. Amer. Soc. Hort. Soc. for* 1936, 34, 199–205.

BURRELL, A. B. (1938). Control of internal cork of apple with boron. *Proc. Amer. Soc. Hort. Soc. for* 1937, 35, 161–75.

BURSTRÖM, H. (1939). Über die Schwermetallkatalyse der Nitrat-Assimilation. *Planta*, 29, 292–305.

BYERS, H. G. (1934). Selenium, vanadium, chromium and arsenic in one soil. *Industr. Engng Chem.* (News ed.), 12, 122.

BYERS, H. G. (1935). Selenium occurrence in certain soils in the United States, with a discussion of related topics. *Tech. Bull. U.S. Dep. Agric.* no. 482, 47 pp.

BYERS, H. G. and KNIGHT, H. G. (1935). Selenium in soils in relation to its presence in vegetation. *Industr. Engng Chem.* 27, 902–4.

CAHILL, V. (1929). Experiments for the control of exanthema in Japanese plum trees. *J. Dep. Agric. W. Aust.* **6**, 388–94.

CALFEE, R. K. and McHARGUE, J. S. (1937). Optical spectroscopic determination of boron. Polarizing attachments. *Industr. Engng Chem.* (Anal. ed.), **9**, 288–90.

CALLAN, R. and HENDERSON, J. A. R. (1929). A new reagent for the colorimetric determination of minute amounts of copper. *Analyst*, **54**, 650–3.

CAMP, A. F. (1945). Zinc as a nutrient in plant growth. *Soil Sci.* **60**, 156–64.

CARNE, W. M. and MARTIN, D. (1937). Preliminary experiments in Tasmania on the relation of internal cork of apples and cork of pears to boron deficiency. *Aust. J. Coun. Sci. Industr. Res.* **10**, 47–56.

CHANDLER, F. B. (1941). Mineral nutrition of the genus *Brassica* with particular reference to boron. *Bull. Maine Agric. Exp. Sta.* no. 404.

CHANDLER, W. H. (1937). Zinc as a nutrient for plants. *Bot. Gaz.* **98**, 625–46.

CHANDLER, W. H., HOAGLAND, D. R. and HIBBARD, P. L. (1932). Little-leaf or rosette in fruit trees. *Proc. Amer. Soc. Hort. Sci.* 1931, **28**, 556–60.

CHANDLER, W. H., HOAGLAND, D. R. and HIBBARD, P. L. (1933). Little-leaf or rosette of fruit trees. II. *Proc. Amer. Soc. Hort. Sci.* 1932, **29**, 255–63.

CHANDLER, W. H., HOAGLAND, D. R. and HIBBARD, P. L. (1934). Little-leaf or rosette of fruit trees. III. *Proc. Amer. Soc. Hort. Sci.* 1933, **30**, 70–86.

CHANDLER, W. H., HOAGLAND, D. R. and HIBBARD, P. L. (1935). Little-leaf or rosette of fruit trees. *Proc. Amer. Soc. Hort. Sci.* 1934, **32**, 11–19.

CHAPMAN, H. D. and VANSELOW, A. P. (1956). In *Citrus Leaves*, **86**, 10–12, 26, 28. Cited by W. Reuter, T. W. Embleton and W. W. Jones (1958). Mineral nutrition of tree crops. *Ann. Rev. Pl. Physiol.* **9**, 175–200.

CHESTERS, C. G. C. and ROLINSON, G. N. (1950). Role of zinc in metabolism. *Nature, Lond.*, **165**, 851–2.

COLEMAN, D. R. K. and GILBERT, F. C. (1939). Manganese and caffeine content of some teas and coffees. *Analyst*, **64**, 726–30.

COLLANDER, R. (1941). Selective absorption of cations by higher plants. *Plant Physiol.* **16**, 691–720.

COLWELL, W. E. and LINCOLN, C. (1942). A comparison of boron deficiency symptoms and potato leafhopper injury on alfalfa. *J. Amer. Soc. Agron.* **34**, 495–8.

CONNOR, J., SCHIMP, N. F. and TEDROW, J. C. F. (1957). A spectrographic study of the distribution of trace elements in some podzolic soils. *Soil Sci.* **83**, 65–73.

COOK, J. W. (1941). Rapid method for determination of manganese in feeds. *Industr. Engng Chem.* (Anal. ed.), **13**, 48–50.

COOK, R. L. and MILLER, C. E. (1939). Some soil factors affecting boron availability. *Proc. Soil Sci. Soc. Amer.* **4**, 297–301.

CORNER, H. H. and SMITH, A. M. (1938). The influence of cobalt on pine disease in sheep. *Biochem. J.* **32**, 1800–5.

COWLING, H. and MILLER, E. J. (1941). Determination of small amounts of zinc in plant materials. *Industr. Engng Chem.* (Anal. ed.), **13**, 145–9.

CRIPPS, E. G. (1956). Boron nutrition of the hop. *J. Hort. Sci.* **31**, 25–34.

DAVIDSON, ANNIE M. M. and MITCHELL, R. L. (1940). The determination of cobalt and chromium in soils. *J. Soc. Chem. Ind., Lond.*, **59**, 232–5.

DAVIS, A. R., MARLOTH, R. H. and BISHOP, C. J. (1928). The inorganic nutrition of the fungi. I. The relation of calcium and boron to growth and spore formation. *Phytopathology*, **18**, 949.

DAVIS, G. K. (1958). Mechanisms of trace element function. *Soil Sci.* **85**, 59–62.

DEAN, L. A. and TRUOG, E. (1935). Determination of manganese and magnesium in soils and silicate rocks. *Industr. Engng Chem.* (Anal. ed.), **7**, 383–5.

DEARBORN, C. H. (1942). Boron nutrition of cauliflower in relation to browning. *Bull. Cornell Univ. Agric. Exp. Sta.* no. 778.

DEARBORN, C. H. and RALEIGH, G. J. (1936). A preliminary note on the control of internal browning of cauliflower by the use of boron. *Proc. Amer. Soc. Hort. Sci.* 1935, **33**, 622–3.

DEARBORN, C. H., THOMPSON, H. C. and RALEIGH, G. J. (1937). Cauliflower browning resulting from a deficiency of boron. *Proc. Amer. Soc. Hort. Sci.* 1936, **34**, 483–7.

DEMAREE, J. B., FOWLER, E. D. and CRANE, H. L. (1933). Report of progress on experiments to control pecan rosette. *Nation. Pecan Assoc. Bull. Proc. Ann. Conv.* **32**, 90–9. (Summary in *Biol. Abstr.* **9**, 1225, 1935.)

DENNIS, A. C. and DENNIS, R. W. G. (1939). Boron and plant life. III. Developments in agriculture and horticulture. *Fertil. Feed. St. J.*, 19 pp. Feb., Mar. and Apr.

DENNIS, A. C. and DENNIS, R. W. G. (1941). Boron and plant life. IV. Developments in agriculture and horticulture, 1939–40. *Fertil. Feed. St. J.* 24 pp. Nov. 1940–Feb. 1941.

DENNIS, A. C. and DENNIS, R. W. G. (1943). Boron and plant life. V. Developments in agriculture and horticulture, 1940–42. *Fertil. Feed. St. J.* 38 pp. in reprint Mar.; Apr.; May.

DENNIS, R. W. G. (1937). The relation of boron to plant growth. *Sci. Prog.* **32**, 58–69.

DENNIS, R. W. G. (1937). Boron and plant life. II. Recent developments in agriculture and horticulture. *Fertil. Feed. St. J.* 15 pp., Sept.–Oct.

DENNIS, R. W. G. and O'BRIEN, D. G. (1937). Boron in agriculture. *Res. Bull. W. Scot. Agric. Coll.* no. 5.

DE ROSE, H. R., EISENMENGER, W. S. and RITCHIE, W. S. (1938). The comparative nutritive effects of copper, zinc, chromium and molybdenum. *Bull. Mass. Agric. Exp. Sta.* 1937, no. 347, pp. 18–19.

DIBLE, W. T., TRUOG, E. and BERGER, K. C. (1954). Boron determination in soils and plants. *Anal. Chem.* **26**, 418–21.
DICK, A. T. (1956). Molybdenum in animal nutrition. *Soil Sci.* **81**, 229–36.
DRAIZE, J. H. and BEATH, O. A. (1935). Observations on the pathology of blind staggers and alkali disease. *Amer. Vet. Med. Ass. J.* **86**, (N.S. **39**), 753–63.
DRAKE, M., SIELING, D. H. and SCARSETH, G. D. (1941). Calcium-boron ratio as an important factor in controlling boron starvation. *J. Amer. Soc. Agron.* **33**, 454–62.
DREGNE, H. E. and POWERS, W. L. (1942). Boron fertilization of alfalfa and other legumes in Oregon. *J. Amer. Soc. Agron.* **34**, 902–12.
DUFRÉNOY, J. and REED, H. S. (1934). Pathological effects of the deficiency or excess of certain ions on the leaves of *Citrus* plants. *Ann. Agron.* N.S. **4**, 637–53.
DUFRÉNOY, J. and REED, H. S. (1942). Coacervates in physical and biological systems. *Phytopathology*, **32**, 568–79.
DUNLOP, G. (1939). *Mineral Deficiencies in Live Stock on British Pastures*. Glasgow: The Scottish Agric. Publ. Co.
DUNLOP, G., INNES, J. R. M., SHEARER, G. D. and WELLS, H. (1939). 'Swayback' studies in North Derbyshire. I. The feeding of copper to pregnant ewes in the control of 'Swayback'. *J. Comp. Path.* **52**, 259–65.
DUNN, F. J. and DAWSON, C. R. (1951). On the nature of ascorbic acid oxidase. *J. Biol. Chem.* **189**, 485–97.
DUNNE, T. C. (1938). 'Wither-tip' or 'Summer Dieback'. *J. Agric. W. Aust.* **15** (2nd Ser.), 120–6.

EARLEY, E. B. (1943). Minor element studies with soybeans. I. Varietal reaction to concentrations of zinc in excess of the nutritional requirement. *J. Amer. Soc. Agron.* **35**, 1012–23.
EATON, F. M. (1935). Boron in soils and irrigation waters and its effect on plants. *Tech. Bull. U.S. Dep. Agric.* no. 448.
EATON, F. M. (1942). Toxicity and accumulation of chloride and sulfate in plants. *J. Agric. Res.* **64**, 357–99.
EATON, S. V. (1940). Effects of boron deficiency and excess on plants. *Plant Physiol.* **15**, 95–107.
EDEN, A. and GREEN, H. H. (1940). Micro-determination of copper in biological material. *Biochem. J.* **34**, 1202–8.
EISLER, B., ROSDAHL, K. G. and THEORELL, H. (1936). Über die Mikrobestimmung des Kupfers mit Hilfe der lichtelektrischen Photometrie. *Biochem. Z.* **285**, 76–7.
ELLIOTT, W. H. (1951). Studies on the enzymic synthesis of glutamine. *Biochem. J.* **49**, 106–12.
ELLIOTT, W. H. (1953). Isolation of glutamine synthetase and glutamotransferase from green peas. *J. Biol. Chem.* **201**, 661–72.
ELTINGE, E. T. (1936). Effect of boron deficiency upon the structure of *Zea mays*. *Plant Physiol.* **11**, 765–78.

ELTINGE, E. T. and REED, H. S. (1940). The effect of zinc deficiency upon the root of *Lycopersicum esculentum*. *Amer. J. Bot.* **27**, 331–5.

EMERSON, R. and LEWIS, C. M. (1939). Factors influencing the efficiency of photosynthesis. *Amer. J. Bot.* **26**, 808–22.

ERKAMA, J. (1947). Über die Rolle von Kupfer und Mangan im Leben der höheren Pflanzen. *Ann. Acad. Sci. Fenn.* A, II, **25**, 1–105.

ERKAMA, J. (1950). On the effect of copper and manganese on the iron status of higher plants. In 'Trace Elements in Plant Physiology'. *Lotsya*, **3**, 53–62.

EVANS, C. E., LATHWELL, D. J. and MEDERSKI, H. J. (1950). Effect of deficient or toxic levels of nutrients in solution on foliar symptoms and mineral content of soybean leaves as measured by spectroscopic methods. *Agron. J.* **42**, 25–32.

EVANS, H. J. and PURVIS, E. R. (1951). Molybdenum status of some New Jersey soils with respect to alfalfa production. *Agron. J.* **43**, 70–1.

EVANS, H. J., PURVIS, E. R. and BEAR, F. E. (1950). Molybdenum nutrition of alfalfa. *Plant Physiol.* **25**, 555–66.

FERGUSON, W. S. (1943). The teart pastures of Somerset. IV. The effect of continuous administration of copper sulphate to dairy cows. *J. Agric. Sci.* **33**, 116–18.

FERGUSON, W. S., LEWIS, A. H. and WATSON, S. J. (1943). The teart pastures of Somerset. I. The cause and cure of teartness. *J. Agric. Sci.* **33**, 44–51.

FILMER, J. F. and UNDERWOOD, E. J. (1937). Enzootic marasmus. Further data concerning the potency of cobalt as a curative and prophylactic agent. *Aust. Vet. J.* **13**, 57–64.

FINCH, A. H. (1933). Pecan rosette, a physiological disease apparently susceptible to treatment with zinc. *Proc. Amer. Hort. Sci.* 1932, **29**, 264–6.

FINCH, A. H. and KINNISON, A. F. (1933). Pecan rosette: soil, chemical and physiological studies. *Tech. Bull. Arizona Agric. Exp. Sta.* no. 47, pp. 407–42.

FISHER, P. L. (1935). Responses of the tomato in solution cultures with deficiencies and excesses of certain essential elements. *Bull. Md Agric. Exp. Sta.* no. 375, pp. 282–98.

FLOYD, B. F. (1908). Leaf spotting of *Citrus*. *Ann. Rep. Florida Agric. Exp. Sta.* no. 91.

FLOYD, B. F. (1917). Dieback, or exanthema of *Citrus* trees. *Bull. Fa Agric. Exp. Sta.* no. 140, 31 pp.

FOSTER, J. S. and HORTON, C. A. (1937). Quantitative spectrographic analysis of biological material. II. *Proc. Roy. Soc.* B, **123**, 422–30.

FOSTER, J. W. (1939). The heavy metal nutrition of fungi. *Bot. Rev.* **5**, 207–39.

FOSTER, J. W. and DENISON, F. W. (1950). Role of zinc in metabolism. *Nature, Lond.*, **166**, 833–4.

FOSTER, J. W. and WAKSMAN, S. A. (1939). The specific effect of zinc and other heavy metals on growth and fumaric-acid production by *Rhizopus*. *J. Bact.* **37**, 599–617.

FRANKE, K. W. and MOXON, A. L. (1936). A comparison of the minimum fatal dose of selenium, tellurium, arsenic and vanadium. *J. Pharmacol.* 58, 454–9.

FREY-WYSSLING, A. (1935). Die unentbehrlichen Elemente der Pflanzennahrung. *Naturwissenschaften,* 23, 767–9.

FUJIMOTO, C. K. and SHERMAN, G. D. (1948). Manganese availability as influenced by steam sterilization of soils. *J. Amer. Soc. Agron.* 40, 527–34.

FUJIMOTO, C. K. and SHERMAN, G. D. (1950). Cobalt content of typical soils and plants of the Hawaiian islands. *Agron. J.* 42, 577–81.

FUJIMOTO, C. K. and SHERMAN, G. D. (1951). Molybdenum content of typical soils and plants of the Hawaiian islands. *Agron. J.* 43, 424–9.

GALLAGHER, P. H. and WALSH, T. (1943). The susceptibility of cereal varieties to manganese deficiency. *J. Agric. Sci.* 33, 197–203.

GAUCH, H. G. and DUGGAR, W. M. (1953). The role of boron in the translocation of sucrose. *Plant Physiol.* 28, 457–67.

GAUCH, H. G. and DUGGAR, W. M. (1954). The physiological action of boron in higher plants: a review and interpretation. *Bull. Univ. Maryland Agric. Exp. Sta.* no. A-80 (Technical), 43 pp.

GEIGEL, A. R. (1935). Effect of boron on the growth of certain green plants. *J. Agric. Univ. Puerto Rico,* 19, 5–28.

GERRETSEN, F. C. (1937). Manganese deficiency of oats and its relation to soil bacteria. *Ann. Bot., Lond.,* N.S. 1, 207–30.

GILBERT, B. E. and McLEAN, F. T. (1928). A 'deficiency disease': The lack of available manganese in a lime-induced chlorosis. *Soil Sci.* 26, 27–31.

GILE, P. L. (1916). Chlorosis of pineapples induced by manganese and carbonate of lime. *Science,* 44, 855–7.

GISIGER, L. (1950). Deficiencies of minor elements caused by excesses. In 'Trace elements in plant physiology'. *Lotsya,* 3, 19–30.

GLASSCOCK, H. H. and WAIN, R. L. (1940). Distribution of manganese in the pea seed in relation to marsh spot. *J. Agric. Sci.* 30, 132–40.

GOLLMICK, F. (1936). Der Einfluss von Zink, Eisen und Kupfer und deren Kombination auf das Wachstum von *Aspergillus niger.* *Z. Bakt.* II, 93, 421–42.

GRAM, E. (1936). Bormangel og nogle andre mangelsygdomme. *Tidsskr. Planteavl.* 41, 401–49.

GREENWOOD, M. and HAYFRON, R. J. (1951). Iron and zinc deficiencies in cacao in the Gold Coast. *Emp. J. Exp. Agric.* 19, 73–86.

GREIG, J. R., DRYERRE, H., GODDEN, W., CRICHTON, A. and OGG, W. G. (1933). Pine: a disease affecting sheep and young cattle. *Vet. J.* 89, 99–110.

GRIGGS, MARY A., JOHNSTIN, RUTH and ELLEDGE, BONNIE E. (1941). Mineral analysis of biological materials. *Industr. Engng Chem.* (Anal. ed.), 13, 99–101.

GRIMM, P. W. and ALLEN, P. J. (1954). Promotion by zinc of the formation of cytochromes in *Ustilago sphaerogena.* *Plant Physiol.* 29, 369–77.

GRIZZARD, A. L. and MATHEWS, E. M. (1942). The effect of boron on seed production of alfalfa. *J. Amer. Soc. Agron.* **34**, 365–8.

GÜSSOW, H. T. (1934). Brown-heart of swede turnips. *Rep. Third Imp. Mycol. Conference*, p. 26.

HAAS, A. R. C. (1932). Some nutritional aspects in mottle-leaf and other physiological diseases of *Citrus*. *Hilgardia*, **6**, 484–559.

HAAS, A. R. C. (1933). Injurious effects of manganese and iron deficiencies on the growth of *Citrus*. *Hilgardia*, **7**, 181–206.

HAAS, A. R. C. (1937). Zinc relation in mottle-leaf of *Citrus*. *Bot. Gaz.* **98**, 65–86.

HAAS, A. R. C. and KLOTZ, L. J. (1931). Some anatomical and physiological changes in *Citrus* produced by boron deficiency. *Hilgardia*, **5**, 175–97.

HAAS, A. R. C. and QUAYLE, H. J. (1935). Copper content of *Citrus* leaves and fruit in relation to exanthema and fumigation injury. *Hilgardia*, **9**, 143–77.

HAMMETT, L. P. and SOTTERY, C. T. (1925). A new reagent for aluminium. *J. Amer. Chem. Soc.* **47**, 142–3.

HAMMOND, W. H. (1928). A rapid method for the detection of zinc in the presence of iron. *Chem. Anal.* **17**, 14.

HARVEY, H. W. (1939). Substances controlling the growth of a diatom. *J. Mar. Biol. Ass. U.K.* **23**, 499–520.

HASELHOFF, E. (1913). Über die Einwirkung von Borverbindungen auf das Pflanzenwachstum. *Landw. Versuchs-Stat.* **79/80**, 399–429.

HATCHER, J. T. and WILCOX, L. V. (1950). Colorimetric determination of boron using carmine. *Anal. Chem.* **22**, 567–9.

HAYWOOD, F. W. and WOOD, A. A. R. (1944). *Metallurgical Analysis by means of the Spekker Photo-electric Absorptiometer.* London.

HEGGENESS, H. G. (1942). Effect of boron applications on the incidence of rust on flax. *Plant Physiol.* **17**, 143–4.

HEINICKE, A. J., REUTHER, W. and CAIN, J. C. (1942). Influence of boron application on preharvest drop of McIntosh apples. *Proc. Amer. Soc. Hort. Sci.* **40**, 31–4.

HEINTZE, S. G. (1938). Readily soluble manganese of soils and marsh spot of peas. *J. Agric. Sci.* **28**, 175–86.

HEINTZE, S. G. (1946). Manganese deficiency in peas and other crops in relation to the availability of soil manganese. *J. Agric. Sci.* **36**, 227–38.

HEINTZE, S. G. (1956). The effect of various soil treatments on the occurrence of marsh spot in peas and on manganese uptake and yield of oats and barley. *Plant and Soil*, **7**, 218–36.

HEWITT, E. J. (1945). 'Marsh spot' in beans. *Nature, Lond.*, **155**, 22–3.

HEWITT, E. J. (1948a). Relation of manganese and some other metals to the iron status of plants. *Nature, Lond.*, **161**, 489.

HEWITT, E. J. (1948b). Experiments on iron metabolism in plants. *Ann. Rep. Agric. Hort. Res. Sta., Long Ashton*, pp. 68–76.

HEWITT, E. J. (1952). Sand and water culture methods used in the study of plant nutrition. *Publications Commonwealth Bur. Hort. and Plantation Crops*, no. 22. Farnham Royal.

HEWITT, E. J. (1953). Metal interrelationships in plant nutrition. I. Effects of some metal toxicities on sugar beet, tomato, oat, potato and marrowstem kale grown in sand culture. *J. Exp. Bot.* **4**, 59–64.

HEWITT, E. J. (1956). Symptoms of molybdenum deficiencies in plants. *Soil Sci.* **81**, 159–71.

HEWITT, E. J., ARGAWALA, S. C. and JONES, E. W. (1950). Effect of molybdenum status on the ascorbic acid content of plants in sand culture. *Nature, Lond.*, **166**, 1119–20.

HEWITT, E. J. and BOLLE-JONES, E. W. (1952a). Molybdenum as a plant nutrient. I. The influence of molybdenum on the growth of some *Brassica* crops in sand culture. *J. Hort. Sci.* **27**, 245–56.

HEWITT, E. J. and BOLLE-JONES, E. W. (1952b). Molybdenum as a plant nutrient. II. Effect of molybdenum deficiency on some *Brassica* crops. *J. Hort. Sci.* **27**, 257–65.

HEWITT, E. J. and HALLAS, D. G. (1951). The use of *Aspergillus niger* (van Tiegh) M strain as a test organism in the study of molybdenum as a plant nutrient. *Plant and Soil*, **3**, 366–408.

HEWITT, E. J. and JONES, E. W. (1947). The production of molybdenum deficiency in plants in sand culture with special reference to tomato and *Brassica* crops. *J. Pomol.* **23**, 254–62.

HEWITT, E. J., JONES, E. W. and WILLIAMS, A. H. (1949). Relation of molybdenum and manganese to the free amino acid content of the cauliflower. *Nature, Lond.*, **163**, 681–2.

HEWITT, E. J. and McCREADY, G. C. (1954). Relation of nitrogen supply to the molybdenum requirement of tomato plants grown in sand culture. *Nature, Lond.*, **174**, 186–7.

HILL, H. and GRANT, E. P. (1935). The growth of turnips in artificial culture. *Sci. Agric.* **15**, 652–9.

HOAGLAND, D. R. (1941). Water culture experiments on molybdenum and copper deficiencies of fruit trees. *Proc. Amer. Soc. Hort. Sci.* **38**, 8–12.

HOAGLAND, D. R. and ARNON, D. I. (1938). The water-culture method for growing plants without soil. *Circ. Univ. Calif. Coll. Agric.* no. 347.

HOAGLAND, D. R., CHANDLER, W. H. and HIBBARD, P. L. (1936). Little-leaf or rosette of fruit trees. V. Effect of zinc on the growth of plants of various types in controlled soil and water culture experiments. *Proc. Amer. Soc. Hort. Sci.* 1935, **33**, 131–41.

HOAGLAND, D. R., CHANDLER, W. H. and STOUT, P. R. (1937). Little-leaf or rosette of fruit trees. VI. Further experiments bearing on the cause of the disease. *Proc. Amer. Soc. Hort. Sci.* 1936, **34**, 210–12.

HOGG, J. (The Ettrick Shepherd) (1831). Remarks on certain diseases of sheep. *Quart. J. Agric.* **2**, 697–706 (and note by the Editor, 706–12).

HOLLAND, E. B. and RITCHIE, W. S. (1939). Report on zinc. *J. Ass. Off. Agric. Chem.* **22**, 333–8.

HOLLEY, K. T. and DULIN, T. G. (1937). A study of ammonia and nitrate nitrogen for cotton. IV. Influence of boron concentration. *Bull. Ga Agric. Exp. Sta.* no. 197.

HOLM-HANSEN, O., GERLOFF, G. C. and SKOOG, F. (1954). Cobalt as an essential element for blue-green algae. *Physiol. Plant.* **7**, 665–75.

HOPKINS, E. F. (1930a). The necessity and function of manganese in the growth of *Chlorella* sp. *Science*, **72**, 609–10.

HOPKINS, E. F. (1930b). Manganese an essential element for a green alga. *Amer. J. Bot.* **17**, 1047.

HOPKINS, E. F. (1934). Manganese an essential element for green plants. *Mem. Cornell Agric. Exp. Sta.* no. 151.

HOPKINS, E. F. and WANN, F. B. (1927). Iron requirement for *Chlorella*. *Bot. Gaz.* **84**, 407–27.

HORNER, C. K., BURK, D., ALLISON, F. E. and SHERMAN, M. S. (1942). Nitrogen fixation by *Azotobacter* as influenced by molybdenum and vanadium. *J. Agric. Res.* **65**, 173–93.

HOTTER, E. (1890). Über das Vorkommen des Bors im Pflanzenreich und dessen physiologische Bedeutung. *Landw. Versuchs-Stat.* **37**, 437–55.

HUMPHREYS, T. E. (1955). An enzyme system from wheat germ catalysing the aerobic oxidation of reduced triphosphopyridine nucleotide (TPN). *Plant Physiol.* **30**, 46–54.

HUNTER, J. G. (1953). The composition of bracken: some major- and trace-element constituents. *J. Sci. Fd Agric.* **4**, 10–20.

HURD-KARRER, ANNIE M. (1934). Selenium injury to wheat plants and its inhibition by sulphur. *J. Agric. Res.* **49**, 343–57.

HURD-KARRER, ANNIE M. (1935). Factors affecting the absorption of selenium from soils by plants. *J. Agric. Res.* **50**, 413–27.

HURD-KARRER, ANNIE M. (1937). Selenium absorption by crops as related to their sulphur requirement. *J. Agric. Res.* **54**, 601–8.

HURD-KARRER, ANNIE M. and KENNEDY, MARY H. (1936). Inhibiting effect of sulphur in selenized soil on toxicity of wheat to rats. *J. Agric. Res.* **52**, 933–42.

HURST, R. R. and MacLEOD, D. J. (1936). Turnip brown heart. *Sci. Agric.* **17**, 209–14.

ICHIOKA, P. S. and ARNON, D. I. (1955). Molybdenum in relation to nitrogen metabolism. II. Assimilation of ammonia and urea without molybdenum by Scenedesmus. *Physiol. Plant.* **8**, 555–60.

ISAAC, W. E. (1934). Researches on the chlorosis of deciduous fruit trees. II. Experiments on chlorosis of peach trees. *Trans. Roy. Soc. S. Afr.* **22**, 187–204.

JACKS, G. V. and SCHERBATOFF, H. (1934). Soil deficiencies and plant diseases. *Tech. Comm. Imp. Bur. Soil Sci.* **31**. Harpenden.

JACKSON, M. L. (1958). *Soil Chemical Analysis*. Englewood Cliffs, N.J.

JAMALAINEN, E. A. (1935a). Tutkimuksia lantun ruskotandista. *Valtion Maatalouskoetoiminnan julkaisuja*, no. 72, pp. 107–16 (with German summary). Cited from Dennis and O'Brien (1937).

JAMALAINEN, E. A. (1935b). Der Einfluss steigender Borsäuremengen auf die Kohlrübenernte. *J. Agric. Soc. Finland*, **7**, 182-6. Cited from Dennis and O'Brien (1937).

JAMES, W. O. (1953). The terminal oxidase in the respiration of the embryo and young roots in barley. *Proc. Roy. Soc. B*, **141**, 280–99.

JAMES, W. O., ROBERTS, E. A. H., BEEVERS, H. and DE KOCK, P. C. (1948). The secondary oxidation of amino acids by the catechol oxidase of Belladonna. *Biochem. J.* **43**, 626–36.

JOHNSON, C. M., STOUT, P. R., BROYER, T. C. and CARLTON, A. B. (1957). Comparative chlorine requirements of different plant species. *Plant and Soil*, **8**, 337–53.

JOHNSTON, E. S. and DORE, W. H. (1929). The influence of boron on the chemical composition and growth of the tomato plant. *Plant Physiol.* **4**, 31–62.

JOHNSTON, J. C. (1933). Zinc sulfate promising new treatment for mottle leaf. *Calif. Citrograph*, **18**, 107, 116–18.

JOLIVETTE, J. P. and WALKER, J. C. (1943). Effect of boron deficiency on the histology of garden beet and cabbage. *J. Agric. Res.* **66**, 167–82.

JONES, H. E. and SCARSETH, G. D. (1944). The calcium-boron balance in plants as related to boron needs. *Soil Sci.* **56**, 15–24.

JONES, L. H., SHEPARDSON, W. B. and PETERS, C. A. (1949). The function of manganese in the assimilation of nitrates. *Plant Physiol.* **24**, 300–6.

JONES, L. H. and THURMAN, D. A. (1957). The determination of aluminium in soil, ash and plant materials using erichrome cyanine RA. *Plant and Soil*, **9**, 131–42.

KAHLENBERG, L. and TRUE, R. H. (1896). On the toxic action of dissolved salts and their electrolytic dissociation. *Bot. Gaz.* **22**, 81–124.

KEILIN, D. and MANN, T. (1938). Polyphenol oxidase. Purification, nature and properties. *Proc. Roy. Soc.* B, **125**, 187–204.

KEILIN, D. and MANN, T. (1939). Laccase, a blue copper protein oxidase from the latex of *Rhus succedanea*. *Nature, Lond.*, **143**, 23–4.

KEILIN, D. and MANN, T. (1940a). Carbonic anhydrase. Purification and nature of the enzyme. *Biochem. J.* **34**, 1163–76.

KEILIN, D. and MANN, T. (1940b). Some properties of laccase from the latex of lacquer trees. *Nature, Lond.*, **145**, 304.

KELLEY, W. P. (1909). The influence of manganese on the growth of pineapples. *Bull. Hawaii Agric. Exp. Sta.* no. 23.

KELLEY, W. P. (1912). The function and distribution of manganese in plants and soils. *Bull. Hawaii Agric. Exp. Sta.* no. 26.

KESSELL, S. L. and STOATE, T. N. (1936). Plant nutrients and pine growth. *Aust. For.* **1**, 4–13.

KESSELL, S. L. and STOATE, T. N. (1938). Pine nutrition. *Bull. W. Aust. For. Dep.* no. 50.

KIDSON, E. B. (1954). Molybdenum content of Nelson soils. *N.Z. J. Sci. Tech.* **36**, 38–45.

KIDSON, E. B. and ASKEW, H. O. (1940). A critical examination of the nitroso-R-salt method for the determination of cobalt in pastures. *N.Z. J. Sci. Tech.* **21** B, 178B–189B.

KIDSON, E. B., ASKEW, H. O. and DIXON, J. K. (1936). Colorimetric determination of cobalt in soils and animal organs. *N.Z. J. Sci. Tech.* **18**, 601–7.

KIDSON, E. B. and MAUNSELL, P. W. (1939). The effect of cobalt compounds on the cobalt content of supplementary fodder crops. *N.Z. J. Sci. Tech.* 21 A, 125 A–128 A.

KLEIN, R. M. (1951). The relation of gas exchange and tyrosinase activity of tomato tissues to the level of boron nutrition of the plants. *Arch. Biochem.* 30, 207–14.

KNIGHT, H. G. (1935). The selenium problem. *J. Ass. Off. Agric. Chem.* 18, 103–8.

KNOP, W. (1860). Ueber die Ernährung der Pflanzen durch wässerige Lösungen bei Ausschluss des Bodens. *Landw. Versuchs-Stat.* 2, 65–99, 270–93.

KNOP, W. (1884). Ueber die Aufnahme verschiedener Substanzen durch die Pflanze, welche nicht zu den Nährstoffen gehören. *Jahresber. Agrik. Chem.* 7, 138–40.

KOLTHOFF, I. M. and LINGANE, J. J. (1952). *Polarography.* Second ed. New York.

KRATZ, W. A. and MYERS, J. (1955). Nutrition and growth of several blue-green algae. *Amer. J. Bot.* 42, 282–7.

KRAUSCH, C. (1882). Ueber Pflanzenvergiftungen. *J. Landw.* 30, 271–91.

KUBOWITZ, F. (1937). Über die chemische Zusammensetzung der Kartoffeloxydase. *Biochem. Z.* 292, 221–9.

KUYPER, J. (1930). Boorzuur tegen de topziekte van de tabak. *Deli Proefstat. te Medan, Sumatra, Vlugschr,* 50, 7 pp.

LARSEN, C. and BAILEY, D. E. (1913). Effect of alkali water on dairy cows. *Bull. S. Dakota Agric. Exp. Sta.* no. 147, pp. 300–25.

LARSEN, C., WHITE, W. and BAILEY, D. E. (1912). Effect of alkali water on dairy products. *Bull. S. Dakota Agric. Exp. Sta.* no. 132, pp. 220–54.

LEDEBOER, M. S. J. (1934). Physiologische onderzoekingen over *Ceratostomella ulmi* (Schwarz) Buisman. Diss. Utrecht.

LEE, H. A. and McHARGUE, J. S. (1928). The effect of a manganese deficiency of the sugar cane plant and its relationship to Pahala blight of sugar cane. *Phytopathology,* 18, 775–86.

LEWIN, J. C. (1954). Silicon metabolism in diatoms. I. Evidence for the role of reduced sulfur compounds in silicon utilization. *J. Gen. Physiol.* 37, 589–99.

LEWIN, J. C. (1955). Silicon metabolism in diatoms. II. Sources of silicon for growth of *Navicula pelliculosa. Plant Physiol.* 30, 129–34.

LEWIS, A. H. (1939). Manganese deficiencies in crops. I. Spraying pea crops with solutions of manganese salts to eliminate marsh spot. *Emp. J. Exp. Agric.* 7, 150–4.

LEWIS, A. H. (1943a). The teart pastures of Somerset. II. Relation between soil and teartness. *J. Agric. Sci.* 33, 52–7.

LEWIS, A. H. (1943b). The teart pastures of Somerset. III. Reducing the teartness of pasture herbage. *J. Agric. Sci.* 33, 58–63.

LEWIS, J. C. (1942). The influence of copper and iodine on the growth of *Azotobacter agile. Amer. J. Bot.* 29, 207–10.

LEWIS, J. C. and POWERS, W. L. (1941). Iodine in relation to plant nutrition. *J. Agric. Res.* 63, 623–37.

214 LIST OF LITERATURE

LIEBIG, G. F., VANSELOW, A. P. and CHAPMAN, H. D. (1943). Effects of gallium and indium on the growth of *Citrus* plants in solution cultures. *Soil Sci.* 56, 173–85.

LINGANE, J. J. and KERLINGER, H. (1941). Polarographic determination of nickel and cobalt. Simultaneous determination in presence of iron, copper, chromium, and manganese, and determination of small amounts of nickel in cobalt compounds. *Industr. Engng Chem.* (Anal. ed.), 13, 77–80.

LIPMAN, C. B. (1938). Importance of silicon, aluminium and chlorine for higher plants. *Soil Sci.* 45, 189–98.

LIPMAN, C. B. and MACKINNEY, G. (1931). Proof of the essential nature of copper for higher plants. *Plant Physiol.* 6, 593–9, 1931.

LOCKWOOD, L. B. (1933). A study of the physiology of *Penicillium Javanicum* Van Beikma with special reference to the production of fat. *Catholic Univ. Amer. Biol. Ser.* 13.

LÖHNIS, M. (1936). Wat Veroorzaakt Kwade Harten in Erwten? *Tijdschr. PlZiekt.* 42, 159–67 (with English summary).

LÖHNIS, M. P. (1950). Injury through excess of manganese. In 'Trace Elements in Plant Physiology'. *Lotsya*, 3, 63–76.

LÖHNIS, M. P. (1951). Manganese toxicity in field and market crops. *Plant and Soil*, 3, 193–222.

LOUNAMAA, J. (1956). Trace elements in plants growing wild on different rocks in Finland. A semi-quantitative spectrographic survey. *Ann. Bot. Soc. Zool. Bot. Fenn.* '*Vanamo*', 29, no. 4, 196 pp.

LOVETT-JANISON, P. L. and NELSON, J. M. (1940). Ascorbic acid oxidase from summer crook-neck squash (*C. pepo condensa*). *J. Amer. Chem. Soc.* 62, 1409–12.

LOWENHAUPT, B. (1942). Nutritional effects of boron on growth and development of the sunflower. *Bot. Gaz.* 104, 316–22.

LUNDEGÅRDH, H. (1929). *Die Quantitative Spektralanalyse der Elemente.* Jena.

LUNDEGÅRDH, H. (1932). *Die Nährstoffaufnahme der Pflanze.* Jena.

LUNDEGÅRDH, H. (1934). *Die Quantitative Spektralanalyse der Elemente.* Zweiter Teil. Jena.

LUNDEGÅRDH, H. (1936). On spectral analysis of inorganic elements. *Landboukhogskolano Ann. (Ann. Agric. Coll. Sweden)*, 3, 49–97.

LUNDEGÅRDH, H. (1939). Mangan als Katalysator der Pflanzenatmung. *Planta*, 29, 419–26.

LUNDEGÅRDH, H. (1945). *Die Blattanalyse. Die wissenschaftlichen und praktischen Grundlagen einer pflanzenphysiologischen Methode der Bestimmung des Düngerbedurfnisses des Bodens.* Jena.

LUNDEGÅRDH, H. and PHILIPSON, T. (1938). The spark-in-flame method for spectral analysis. *Landboukhogskolano Ann. (Ann. Agric. Coll. Sweden)*, 5, 249–60.

MACARTHUR, M. (1940). Histology of some physiological disorders of the apple fruit. *Canad. J. Res.* Sect. C, 18, 26–34.

McGEORGE, W. T. (1924). Iron, aluminum and manganese in the soil solution of Hawaiian soils. *Soil Sci.* 18, 1–11.

McHARGUE, J. S. (1922). The role of manganese in plants. *J. Amer. Chem. Soc.* **44**, 1592–8.

McHARGUE, J. S. (1923). Effect of different concentrations of manganese sulphate on the growth of plants in acid and neutral soils and the necessity of manganese as a plant nutrient. *J. Agric. Res.* **24**, 781–94.

McHARGUE, J. S. (1926 a). Manganese and plant growth. *J. Industr. Engng Chem.* **18**, 172.

McHARGUE, J. S. and CALFEE, R. K. (1931 a). Effect of Mn, Cu and Zn on yeast. *Plant Physiol.* **6**, 559–66.

McHARGUE, J. S. and CALFEE, R. K. (1931 b). Effect of Mn, Cu and Zn on growth and metabolism of *Aspergillus flavus* and *Rhizopus nigricans*. *Bot. Gaz.* **91**, 183–93.

McHARGUE, J. S. and CALFEE, R. K. (1932). Determination of boron spectroscopically. *Industr. Engng Chem.* (Anal. ed.), **4**, 385–8.

McILRATH, W. J. and DE BRUYN, J. A. (1956). Calcium–boron relationships in Siberian millet. *Soil Sci.* **81**, 301–10.

McILRATH, W. J. and PALSER, B. F. (1956). Responses of tomato, turnip and cotton to variations in boron nutrition. I. Physiological responses. *Bot. Gaz.* **118**, 43–52.

McLARTY, H. R., WILCOX, J. C. and WOODBRIDGE, C. G. (1937). A yellowing of alfalfa due to boron deficiency. *Sci. Agric.* **17**, 515–17.

McLEAN, R. C. and HUGHES, W. L. (1936). The quantitative distribution of boron in *Vicia faba* and *Gossypium herbaceum*. *Ann. Appl. Biol.* **23**, 231–44.

McMURTREY, J. E. (1929). The effect of boron deficiency on the growth of tobacco plants in aerated and unaerated solutions. *J. Agric. Res.* **38**, 371–80.

McMURTREY, J. E. (1933). Distinctive effects of the deficiency of certain essential elements on the growth of tobacco plants in solution cultures. *Tech. Bull. U.S. Dep. Agric.* no. 340, pp. 1–42.

McMURTREY, J. E. (1935). Boron deficiency in tobacco under field conditions. *J. Amer. Soc. Agron.* **27**, 271–3.

McMURTREY, J. E. and ROBINSON, W. O. (1938). Neglected soil constituents that affect plant and animal development. *Yearb. U.S. Dep. Agric.* pp. 807–29.

McNAUGHT, K. J. (1938). The cobalt content of North Island pastures. *N.Z. J. Sci. Tech.* **20 A**, 14 A–30 A.

MacVICAR, R. and BURRIS, R. H. (1948). Relation of boron to certain plant oxidases. *Arch. Biochem.* **17**, 31–9.

MacVICAR, R. and STRUCKMEYER, B. E. (1946). The relation of photoperiod to the boron requirement of plants. *Bot. Gaz.* **107**, 454–61.

MAGNESS, J. R., DEGMAN, E. S., BATJER, L. P. and REGEIMBAL, L. O. (1937). Effect of nutritional treatments on internal cork of apples. *Proc. Amer. Soc. Hort. Sci.* 1936, **34**, 206–9.

MAJDEL, J. (1930). Universale gravimetrische Methode der Trennung und Bestimmung des Mangans. *Z. anal. Chem.* **81**, 14–26.

MANN, M. (1932). Calcium and magnesium requirements of *Aspergillus niger*. *Bull. Torrey Bot. Cl.* **59**, 443–88.

MARMOY, F. B. (1939). The determination of molybdenum in plant materials. *J. Soc. Chem. Ind., Lond. (Trans.)*, **58**, 275–6.

MARSH, R. P. (1942). Comparative study of the calcium-boron metabolism of representative dicots and monocots. *Soil Sci.* **53**, 75–8.

MARSH, R. P. and SHIVE, J. W. (1941). Boron as a factor in the calcium metabolism of the corn plant. *Soil Sci.* **51**, 141–51.

MARSTON, H. R. and DEWEY, D. W. (1940). The estimation of cobalt in plant and animal tissues. *Aust. J. Exp. Biol. Med. Sci.* **18**, 343–52.

MARTIN, J. P. (1934). Boron deficiency symptoms in sugar cane. *Hawaii. Plant. Res.* **38**, 95–107.

MASCHHAUPT, J. G. (1934). Das Rätsel der Dörrfleckenkrankheit. *Z. PflErnähr. Düng.* **13**, 313–20.

MAZÉ, P. (1914). Influences respectives des éléments de la solution minérale sur le développement du maïs. *Ann. Inst. Pasteur,* **28**, 1–48.

MAZÉ, P. (1915). Détermination des éléments minéraux rares nécessaires au développement du maïs. *C.R. Acad. Sci., Paris,* **160**, 211–14.

MAZÉ, P. (1919). Recherche d'un solution purement minérale capable d'assurer l'évolution complète du maïs cultivé a l'abri des microbes. *Ann. Inst. Pasteur,* **33**, 139–73.

MEDINA, A. and NICHOLAS, D. J. D. (1957). Some properties of a zinc-dependent hexokinase from *Neurospora crassa*. *Biochem. J.* **66**, 573–8.

MEIKLEJOHN, G. T. and STEWART, C. P. (1941). Ascorbic acid oxidase from cucumber. *Biochem. J.* **35**, 755–60.

MELDRUM, N. U. (1934). *Cellular Respiration*. London.

MELVIN, E. H. and O'CONNOR, R. T. (1941). Spectrochemical analysis of trace elements in fertilizers. Boron, manganese and copper. *Industr. Engng Chem.* (Anal. ed.), **13**, 520–4.

MENZEL, R. G. and JACKSON, M. L. (1951). Determination of copper and zinc in soils and plants. *Anal. Chem.* **23**, 1861–3.

METZ, O. (1930). Über Wachstum und Farbstoffbildung einiger Pilze unter dem Einfluss von Eisen, Zink, und Kupfer. *Arch. Mikrobiol.* **1**, 197–251.

MILLER, J. T. and BYERS, H. G. (1937). Selenium in plants in relation to its occurrence in soils. *J. Agric. Res.* **55**, 59–68.

MILLER, W. T. and SCHOENING, H. W. (1938). Toxicity of selenium fed to swine in the form of sodium selenite. *J. Agric. Res.* **56**, 831–42.

MILLER, W. T. and WILLIAMS, K. T. (1940*a*). Minimum lethal dose of selenium, as sodium selenite, for horses, mules, cattle and swine. *J. Agric. Res.* **60**, 163–73.

MILLER, W. T. and WILLIAMS, K. T. (1940*b*). Effect of feeding repeated small doses of selenium as sodium selenite to equines. *J. Agric. Res.* **61**, 353–68.

MILLIKAN, C. R. (1947). Effect of molybdenum on the severity of toxicity symptoms in flax induced by an excess of either manganese, zinc, copper, nickel or cobalt in the nutrient solution. *J. Aust. Inst. Agric. Sci.* **13**, 180–6.

MILLIKAN, C. R. (1948). Antagonism between molybdenum and certain heavy metals in plant nutrition. *Nature, Lond.,* **161**, 528.

MILLIKAN, C. R. (1950). Relation between nitrogen sources and the effects on flax of an excess of manganese or molybdenum in the nutrient solution. *Aust. J. Sci. Res.* B, **3**, 450–73.

MILNER, G. W. C. (1957). *The Principles of and Applications of Polarography and Other Electroanalytical Processes.* New York.

MINARIK, C. E. and SHIVE, J. W. (1939). The effect of boron in the substrate on calcium accumulation by soybean plants. *Amer. J. Bot.* **26**, 827–31.

MITCHELL, R. L. (1936). Spectrographic analysis of soils by the Lundegårdh method. *J. Soc. Chem. Ind., Lond. (Trans.),* **55**, 267–9.

MITCHELL, R. L. (1940). The spectrographic determination of trace elements in soils. I. The cathode layer arc. *J. Soc. Chem. Ind., Lond. (Trans.),* **59**, 210–13.

MITCHELL, R. L. (1941). The spectrographic analysis of solutions by a modified Ramage flame emission method. *J. Soc. Chem. Ind., Lond. (Trans.),* **60**, 95–8.

MITCHELL, R. L. (1944). The distribution of trace elements in soils and grasses. *Proc. Nutr. Soc.* **1**, 183–9.

MITCHELL, R. L. (1948). The spectrographic analysis of soils, plants and related materials. *Tech. Commun. Commonwealth Bur. Soil Sci.* no. 44, 183 pp. Harpenden.

MITCHELL, R. L. (1955). Trace elements. In *Chemistry of Soil,* pp. 253–85. Ed. E. F. Bear. New York.

MITCHELL, R. L. and ROBERTSON, I. M. (1936). The effect of aluminium on the flame spectra of the alkaline earths: a method for the determination of aluminium. *J. Soc. Chem. Ind., Lond. (Trans.),* **55**, 269–72.

MITCHELL, R. L., SCOTT, R. O., STEWART, A. B. and STEWART, J. (1941). Cobalt manuring and pining in stock. *Nature, Lond.,* **148**, 725.

MONK, R. J. (1955). Boron deficiency symptoms in raspberries. *N.Z. J. Sci. Tech.* **36**, 610–13.

MORRIS, A. A. (1938). Effects of boron treatment in the control of 'hard fruit' *Citrus. J. Pom. Hort. Sci.* **16**, 167–81.

MORRIS, H. D. and PIERRE, W. H. (1947). The effect of calcium, phosphorus and iron on the tolerance of Lespedeza to manganese toxicity in culture solutions. *Proc. Soil Sci. Soc. Amer.* **12**, 382–6.

MORRIS, H. D. and PIERRE, W. H. (1949). Minimum concentrations of manganese necessary for injury to various legumes in culture solutions. *Agron. J.* **41**, 107–12.

MOSHER, W. A., SAUNDERS, D. H., KINGERY, L. K. and WILLIAMS, R. J. (1936). Nutritional requirements of the pathogenic mould *Trichophyton interdigitale. Plant Physiol.* **11**, 795–806.

MOWRY, H. and CAMP, A. F. (1934). A preliminary report on zinc sulfate as a corrective for bronzing of tung trees. *Bull. Fa Agric. Exp. Sta.* no. 273, pp. 1–34.

MOXON, A. L. (1937). Alkali disease or selenium poisoning. *Bull. S. Dakota Agric. Exp. Sta.* no. 311, 91 pp.

MUHR, G. R. (1940). Available boron as affected by soil treatments. *Proc. Soil Sci. Soc. Amer.* **5**, 220–6.

218   LIST OF LITERATURE

MUHR, G. R. (1942). Plant symptoms of boron deficiency and the effect of borax on the yield and chemical composition of several crops. *Soil Sci.* **54**, 55–65.

MUIR, W. R. (1936). The teart pastures of Somerset. *Agric. Progr.* **13**, 53–61.

MULDER, E. G. (1948). Importance of molybdenum in the nitrogen metabolism of microorganisms and higher plants. *Plant and Soil*, **1**, 94–119.

MULDER, E. G. (1950). Importance of copper and molybdenum in the nutrition of higher plants and microorganisms. In 'Trace elements in plant physiology'. *Lotsya*, **3**, 41–52.

MULDER, E. G., BAKEMA, K. and VEEN, W. L. VAN (1959). Molybdenum in symbiotic nitrogen fixation and in nitrate assimilation. *Plant and Soil*, **10**, 319–34.

MULDER, E. G., BOXMA, R. and VEEN, W. L. VAN (1959). The effect of molybdenum and nitrogen deficiencies on nitrate reduction in plant tissues. *Plant and Soil*, **10**, 335–55.

MYERS, V. C., MULL, J. W. and MORRISON, D. B. (1928). The estimation of aluminium in animal tissues. *J. Biol. Chem.* **78**, 595–604.

NAFTEL, J. A. (1939). Colorimetric determination of boron. *Industr. Engng Chem.* (Anal. ed.), **11**, 407–9.

NAIR, G. G. K. and MEHTA, B. V. (1959). Status of zinc in soils of Western India. *Soil Sci.* **87**, 155–9.

NASON, A. (1950). Effect of zinc deficiency on the synthesis of tryptophane by *Neurospora* extracts. *Science*, **112**, 111–12.

NASON, A. (1952). Metabolism of micronutrient elements in higher plants. II. Effect of copper deficiency on the isocitric enzyme in tomato leaves. *J. Biol. Chem.* **198**, 643–53.

NASON, A., ABRAHAM, R. G. and AVERBACH, B. C. (1954). The enzymic reduction of nitrite to ammonia by reduced pyridine nucleotides. *Biochim. Biophys. Acta*, **15**, 160–1.

NASON, A., KAPLAN, N. O. and COLOWICK, S. P. (1951). Changes in enzymatic constitution in zinc-deficient *Neurospora*. *J. Biol. Chem.* **188**, 397–406.

NASON, A., KAPLAN, N. O. and OLDEWURTEL, H. A. (1953). Further studies of nutrient conditions affecting enzymatic constitution in zinc-deficient *Neurospora*. *J. Biol. Chem.* **201**, 435–44.

NEISH, A. C. (1939). Studies on chloroplasts. II. Their chemical composition and the distribution of certain metabolites between the chloroplasts and the remainder of the leaf. *Biochem. J.* **33**, 300–8.

NELSON, E. M., HURD-KARRER, A. M. and ROBINSON, W. O. (1933). Selenium as an insecticide. *Science*, **78**, 124.

NELSON, J. M. and DAWSON, C. R. (1944). Tyrosinase. *Adv. Enzymol.* **4**, 99–152.

NEWELL, W., MOWRY, H. and BARNETTE, R. M. (1930). The tung-oil tree. *Bull. Fa Agric. Exp. Sta.* no. 221.

NICHOLAS, D. J. D. (1949). The manganese and iron content of crop plants as determined by chemical methods. *J. Hort. Sci.* **25**, 60–77.

NICHOLAS, D. J. D. (1958). Some biochemical aspects of nitrogen fixation. In *Nutrition of the Legumes*, ed. E. G. Hallsworth. London.

NICHOLAS, D. J. D. and FIELDING, A. H. (1947). The use of *Aspergillus niger* (M strain) in the bioassay of magnesium, copper, zinc and molybdenum in soils. I. *Ann. Rep. Agric. Hort. Res. Sta., Long Ashton*, pp. 126–37.

NICHOLAS, D. J. D. and FIELDING, A. H. (1950). Use of *Aspergillus niger* as a test organism for determining molybdenum available in soils to crop plants. *Nature, Lond.*, **166**, 342–3.

NICHOLAS, D. J. D. and NASON, A. (1955). Role of molybdenum as a constituent of nitrate reductase from soybean leaves. *Plant Physiol.* **30**, 135–8.

NICHOLAS, D. J. D., NASON, A. and McELROY, W. D. (1953). Effect of molybdenum deficiency on nitrate reductase in cell-free extracts of *Neurospora* and *Aspergillus*. *Nature, Lond.*, **172**, 34.

NOBBE, F. and SIEGERT, T. (1862, 1863). Ueber das Chlor als spezifischer Nährstoff der Buchweizenpflanze. *Landw. Versuchs-Stat.* **4**, 318–40; **5**, 116–36.

O'CONNOR, R. T. (1941). Spectrochemical analysis of trace elements in fertilizers. Zinc. *Indust. Engng Chem.* (Anal. ed.), **23**, 597–600.

OGILVIE, N. and HICKMAN, C. J. (1936). Progress report on vegetable diseases. *Ann. Rep. Agric. Hort. Res. Sta., Long Ashton*, pp. 139–46.

OLSEN, C. (1934). The absorption of manganese by plants. *C.R. trav. lab. Carlsberg*, **20**, no. 2, 34 pp.

OLSEN, C. (1936). Absorption of manganese by plants. II. The toxicity of manganese to various plant tissues. *C.R. trav. lab. Carlsberg*, **21**, 124–43.

OLSEN, L. C. and DE TURK, E. E. (1940). Rapid microdetermination of boron by means of quinalizarin and a photoelectric colorimeter. *Soil Sci.* **50**, 257–64.

ORTON, W. A. and RAND, F. V. (1914). Pecan rosette. *J. Agric. Res.* **3**, 149–74.

OSERKOWSKY, J. and THOMAS, H. E. (1933). Exanthema in pears and its relation to copper deficiency. *Science*, **78**, 315–16.

OSTERHOUT, W. J. V. (1908). Die Schutzwirkung des Natriums für Pflanzen. *Jahrb. wiss. Bot.* **46**, 121–36.

OSTERHOUT, W. J. V. (1912). Plants which require sodium. *Bot. Gaz.* **54**, 532–6.

OTTO, R. (1893). Untersuchungen über das Verhalten der Pflanzenwurzeln gegen Kupfersalzlösungen. *Z. PflKrankh*, **3**, 322–34.

OVINGE, A. (1935). Het optreden van kwade harten in Schokkers in Zeeland in 1934. *Landbouw. Tijdschr.* **47**, 375–83.

OVINGE, A. (1938). Kwade Harten-Proeven in Zeeland in 1937. *Tijdschr. PlZiekt.* **44**, 208–13.

OWEN, O. and MASSEY, D. M. (1953). Lime induced manganese deficiency in glasshouse roses. *Plant and Soil*, **5**, 81–6.

220   LIST OF LITERATURE

PALSER, B. F. and McILRATH, W. J. (1956). Responses of tomato, turnip and cotton to variations in boron nutrition. II. Anatomical responses. *Bot. Gaz.* 118, 53–71.

PARKER, E. R. (1934). Experiments on the treatment of mottle-leaf of *Citrus* trees. *Proc. Amer. Soc. Hort. Sci.* 1933, 31, 98–107.

PARKER, E. R. (1936). Experiments on the treatment of mottle-leaf of *Citrus* trees. II. *Proc. Amer. Soc. Hort. Sci.* 1935, 33, 82–6.

PARKER, E. R. (1937). Experiments on the treatment of mottle-leaf of *Citrus* trees. *Proc. Amer. Soc. Hort. Sci.* 1936, 34, 213–15.

PATTANAIK, S. (1950a). The effect of boron on the catalase activity of the rice plant. *Current Sci.* 19, 153–4.

PATTANAIK, S. (1950b). The effect of manganese on the catalase activity of the rice plant. *Plant and Soil.* 2, 418–19.

PERRY, V. G., WEDDELL, W. H. and WRIGHT, E. R. (1950). Multipurpose method of spectrographic analysis. *Anal. Chem.* 22, 1516–18.

PETHYBRIDGE, G. H. (1936). Marsh spot in pea seeds: is it a deficiency disease? *J. Minist. Agric.* 43, 55–8.

PETTINGER, N. A., HENDERSON, R. G. and WINGARD, A. (1932). Some nutritional disorders in corn grown in sand cultures. *Phytopathology,* 22, 33–51.

PFEFFER, W. (1900). *The Physiology of Plants,* vol. 1. English ed. Trans. and ed. by A. J. Ewart. Oxford.

PHILIPSON, T. (1953). Boron in plant and soil, with special regard to Swedish agriculture. *Acta Agric. Scand.* 3, 121–242.

PIPER, C. S. (1938). The occurrence of 'reclamation disease' in cereals in South Australia. *Austr. Coun. Sci. Ind. Res.,* Pamphlet 78, 24–8.

PIPER, C. S. (1940). Molybdenum as an essential element for plant growth. *J. Aust. Inst. Agric. Sci.* 6, 162–4.

PIPER, C. S. (1941). Marsh spot of peas: a manganese deficiency disease. *J. Agric. Sci.* 31, 448–53.

PIPER, C. S. (1942a). Investigations on copper deficiency in plants. *J. Agric. Sci.* 32, 143–78.

PIPER, C. S. (1942b). *Soil and Plant Analysis.* Adelaide.

PITTMAN, H. A. (1936). Exanthema of *Citrus,* Japanese plums and apple trees in Western Australia. *J. Dep. Agric. W. Aust.* Second Ser. 13, 187–93.

PITTMAN, H. A. and OWEN, R. C. (1936). Anthracnose and mottle leaf of *Citrus* in Western Australia. *J. Dep. Agric. W. Aust.* Second Ser. 13, 137–42.

POPP, M., CONTZEN, J. and GERICKE, S. (1934). Das Rätsel der Dörrfleckenkrankheit. *Z. PflErnähr. Düng.* 13, 66–73.

PUGLIESE, A. (1913). Sulla biochimica del manganese; contributo alla conoscenza dei rapporti tra manganese en ferro in relazione alla vegetazione. *Atti Ist. Sci. nat. Napoli,* ser. 6, 10, 285–326.

PURVIS, E. R. and HANNA, W. J. (1940). Vegetable crops affected by boron deficiency in eastern Virginia. *Bull. Va Agric. Exp. Sta.* no. 105.

PURVIS, E. R. and PETERSON, N. K. (1956). Methods of soil and plant analysis for molybdenum. *Soil Sci.* 81, 223–8.

PURVIS, E. R. and RUPRECHT, R. W. (1935). Borax as a fertilizer for celery. *Amer. Fertilizer*, 21 Sept. Cited from Dennis and O'Brien (1937).

PURVIS, E. R. and RUPRECHT, R. W. (1937). Cracked stem of celery caused by a boron deficiency in the soil. *Bull. Fa Agric. Exp. Sta.* no. 307.

RALEIGH, G. J. (1939). Evidence for the essentiality of silicon for growth of the beet plant. *Plant Physiol.* 14, 823–8.

RALEIGH, G. J. (1948). Effects of the sodium and the chloride ion in the nutrition of the table beet in culture solutions. *Proc. Amer. Soc. Hort. Sci.* 51, 433–6.

RAMAMOORTHY, B. and VISWANATH, B. (1946). Comparative studies on Indian soils. Spectroscopic estimation of boron contents. *Indian J. Agric. Sci.* 16, 420–6.

RAULIN, J. (1869). Études chimiques sur la végétation. *Ann. Sci. nat. Bot*, 5 Sér. 11, 93–299.

REED, H. S. (1938). Cytology of leaves affected with little-leaf. *Amer. J. Bot.* 25, 174–86.

REED, H. S. (1939). The relation of copper and zinc salts to leaf structure. *Amer. J. Bot.* 26, 29–33.

REED, H. S. (1941). Effects of zinc deficiency on cells of vegetative buds. *Amer. J. Bot.* 28, 10–17.

REED, H. S. (1942). The relation of zinc to seed production. *J. Agric. Res.* 64, 635–44.

REED, H. S. and DUFRÉNOY, J. (1933). Effets de l'affectation dite 'mottle-leaf' sur la structure cellulaire des *Citrus*. *Rev. gen. bot.* 46, 33–44.

REED, H. S. and DUFRÉNOY, J. (1935). The effects of zinc and iron salt on the cell structure of mottled orange leaves. *Hilgardia*, 9, 113–37.

REED, H. S. and DUFRÉNOY, J. (1942). Catechol aggregates in the vacuoles of cells of zinc-deficient plants. *Amer. J. Bot.* 29, 544–51.

REED, J. F. and CUMMINGS, R. W. (1940). Determination of zinc in plant materials using the dropping mercury electrode. *Industr. Engng Chem.* (Anal. ed.), 12, 489–92.

REED, J. F. and CUMMINGS, R. W. (1941). Determination of copper in plant materials using the dropping mercury electrode. *Industr. Engng Chem.* (Anal. ed.), 13, 124–7.

REEVE, E. and SHIVE, J. W. (1944). Potassium–boron and calcium–boron relationships in plant nutrition. *Soil Sci.* 57, 1–14.

REHM, S. (1937). Der Einfluss der Borsäure auf Wachstum und Salzaufnahme von *Impatiens balsamina*. *Jb. wiss. Bot.* 85, 788–814.

REISENAUER, H. M. (1956). Molybdenum content of alfalfa in relation to deficiency symptoms and responses to molybdenum fertilization. *Soil Sci.* 81, 237–42.

REUTHER, W. and BURROWS, F. W. (1942). The effect of manganese sulfate on the photosynthetic activity of frenched tung foliage. *Proc. Amer. Soc. Hort. Sci.* 40, 73–6.

REUTHER, W. and DICKEY, R. D. (1937). A preliminary report on frenching of tung trees. *Bull. Fa Agric. Exp. Sta.* no. 318, pp. 1–21.

RICHES, J. P. R. (1947). Preliminary experiments on the use of synthetic resins in the estimation of trace elements. *Chem. & Ind.* (*Rev.*), pp. 656–8.

RIPPEL, A. (1923). Über die durch Mangan verursachte Eisenchlorose bei grünen Pflanzen. *Biochem. Z.* 140, 315–23.

ROACH, W. A. (1938). Plant injection for diagnostic and curative purposes. *Tech. Comm. Imp. Bur. Hort. Plant. Crops*, no. 10. E. Malling.

ROACH, W. A. (1939). Plant injection as a physiological method. *Ann. Bot., Lond.*, N.S. 3, 155–226.

ROBERG, M. (1928). Über die Wirkung von Eisen-, Zink-, und Kupfersalzen auf *Aspergillen. Zbl. Bakt.* II, 74, 333–71.

ROBERG, M. (1931). Weitere Untersuchungen über die Bedeutung des Zinks für *Aspergillus niger. Zbl. Bakt.* II, 84, 196–230.

ROBERG, M. (1932). Ein Beitrag zur Stoffwechselphysiologie der Grünalgen. II. Über die Wirkung von Eisen-, Zink- und Kupfersalzen. *Jb. wiss. Bot.* 76, 311–32.

ROBINSON, W. O. (1933). Determination of selenium in wheat and soils. *J. Ass. Off. Agric. Chem.* 16, 423–4.

ROGERS, C. H. (1938). Growth of *Phymatotrichum omnivorum* in solutions with varying amounts of certain mineral elements. *Amer. J. Bot.* 25, 621–4.

ROGERS, L. H. (1935). Spectrographic microdetermination of zinc. Preliminary note. *Industr. Engng Chem.* (Anal. ed.), 7, 421–3.

ROGERS, L. H. and GALL, O. E. (1937). Microdetermination of zinc. Comparison of spectrographic and chemical methods. *Industr. Engng Chem.* (Anal. ed.), 9, 42–4.

ROGERS, L. H. and WU, C. (1948). Zinc uptake by oats as influenced by application of lime and phosphate. *J. Amer. Soc. Agron.* 40, 563–6.

ROWE, E. A. (1936). A study of heart-rot of young sugar beet plants grown in culture solutions. *Ann. Bot. Lond.*, 50, 735–46.

RUBINS, E. J. (1956). Molybdenum deficiencies in the United States. *Soil Sci.* 8, 191–7.

RUSH, E. M. and YOE, J. H. (1954). Colorimetric determination of zinc and copper with 2-carboxy-2′-hydroxy-5′-sulfoformazylbenzene. *Anal. Chem.* 26, 1345–7.

RUSOFF, L. L., ROGERS, L. H. and GADDUM, L. W. (1937). Quantitative determination of copper and estimation of other trace elements by spectrographic methods in wire grasses from 'salt sick' and healthy areas. *J. Agric. Res.* 55, 731–8.

RUSSELL, F. C. (1944). Minerals in pasture deficiencies and excesses in relation to animal health. *Tech. Comm. Imp. Bur. Anim. Nutrition*, no. 15, 91 pp.

SACHS, J. (1860). Ueber die Erziehung von Landpflanzen in Wasser. *Bot. Z.* 18, 113–17.

Sachs, J. (1860, 1861). Vegetationsversuche mit Ausschluss des Bodens über die Nahrstoffe und sonstigen Ernährungsbedingungen von Maïs, Bohnen und anderen Pflanzen. *Landw. Versuchs-Stat.* 2, 219–68; 3, 30–44.

Sakamura, T. (1934). Ammonio- und Nitratophilie bei *Aspergillus oryzae* im besonderen Zusammenhang mit Schwermetallen. *J. Fac. Sci. Hokkaido Imp. Univ.* Ser. v, 3, 121–38.

Sakamura, T. (1936). Über einige für die Kultur von *Aspergillen* notwendigen Schwermetalle und das Befreiungsverfahren der Nährlösung von ihren Spuren. *J. Fac. Sci. Hokkaido Imp. Univ.* Ser. v, 4, 99–116.

Sakamura, T. and Yoshimura, F. (1933). Über die Bedeutung der H-Ionenkonzentration und die wichtige Rolle einiger Schwermetallsalze bei der Kugelzellbildung der *Aspergillen*. *J. Fac. Sci. Hokkaido Imp. Univ.* Ser. v, 2, 317–31.

Samuel, G. and Piper, C. S. (1928). Grey speck (manganese deficiency) disease of oats. *J. Agric. S. Aust.* 31, 696–705, 789–99.

Samuel, G. and Piper, C. S. (1929). Manganese as an essential element for plant growth. *Ann. Appl. Biol.* 16, 493–524.

Sandell, E. B. (1950). *Colorimetric Determination of Traces of Metals.* Second ed. New York.

Scharrer, K. and Schropp, W. (1934). Wasser- und Sandkulturversuche mit Mangan. *Z. Pflanzenernähr., Düng. u. Bodenk.* A, 36, 1–15.

Schimp, N. F., Connor, J., Prince, A. L. and Bear, F. E. (1957). Spectrochemical analysis of soils and biological materials. *Soil Sci.* 83, 51–64.

Schmucker, T. (1933). Zur Blütenbiologie tropischer Nymphaea-Arten (Bor als entscheidender Faktor). *Planta*, 18, 642–50.

Schmucker, T. (1935). Über den Einfluss von Borsäure auf Pflanzen, insbesondere keimende Pollenkörner. *Planta*, 23, 264–83.

Schoening, H. W. (1936). Production of so-called 'alkali disease' in hogs by feeding corn grown in affected area. *North Amer. Vet.* 17, 22–8.

Scholz, W. (1934). Über die Chlorose der blauen Lupine und Serradella in ihrer Beziehung zum Eisen und Mangan. *Z. Pflanzenernähr. Düng. u. Bodenk.* A, 35, 88–101.

Scott, R. O. (1945). The effect of extraneous elements on spectral line intensity in the cathode-layer arc. *J. Soc. Chem. Ind.* 64, 189–94.

Scott, R. O. (1946). The spectrographic determination of trace elements in the cathode-layer arc by the variable internal standard method. *J. Soc. Chem. Ind.* 65, 291–7.

Shive, J. W. (1941). Significant roles of trace elements in the nutrition of plants. *Plant Physiol.* 16, 435–45.

Sideris, C. P. (1937). Colorimetric determination of manganese. *Industr. Engng Chem.* (Anal. ed.), 9, 445–6.

Sideris, C. P. (1940). Improvement of formaldoxime colorimetric method for manganese. *Industr. Engng Chem.* (Anal. ed.), 12, 307.

SIDERIS, C. P. and YOUNG, H. Y. (1949). Growth and chemical composition of *Ananas comosus* (L.) Merr. in solution cultures with different iron–manganese ratios. *Plant Physiol.* 24, 416–40.

SISLER, E. C., DUGGAR, W. M. and GAUCH, H. G. (1956). The role of boron in the translocation of organic compounds in plants. *Plant Physiol.* 31, 11–17.

SJOLLEMA, B. (1933). Kupfermangel als Ursache von Krankheiten bei Pflanzen und Tieren. *Biochem. Z.* 267, 151–6.

SJOLLEMA, B. (1938). Kupfermangel als Ursache von Tierkrankheiten. *Biochem. Z.* 295, 272–376.

SKOK, J. (1941). Effect of boron on growth and development of the radish. *Bot. Gaz.* 103, 280–94.

SKOOG, F. (1940). Relationships between zinc and auxin in the growth of higher plants. *Amer. J. Bot.* 27, 939–51.

SMITH, G. S. (1935). The determination of small amounts of boron by means of quinalizarin. *Analyst*, 60, 735–9.

SMITH, M. E. and BAYLISS, N. S. (1942). The necessity of zinc for *Pinus radiata*. *Plant Physiol.* 17, 303–10.

SMITH, M. I., STOHLMAN, E. F. and LILLIE, R. D. (1937). The toxicity and pathology of selenium. *J. Pharmacol.* 60, 449–70.

SMITH, R. E. and THOMAS, H. E. (1928). Copper sulphate as a remedy for exanthema in prunes, apples, pears and olives. *Phytopathology*, 18, 449–54.

SNYDER, E. and HARMON, F. N. (1942). Some effects of zinc sulphate on the Alexandria grape. *Proc. Amer. Soc. Hort. Sci.* 40, 325–7.

SNYDER, G. B. and DONALDSON, R. W. (1937). The use of borax in controlling dark center of turnips. *Proc. Amer. Soc. Hort. Sci.* 1936, 34, 480–2.

SOMERS, I. I., GILBERT, S. G. and SHIVE, J. W. (1942). The iron–manganese ratio in relation to the respiratory CO_2 and deficiency-toxicity symptoms in soybeans. *Plant Physiol.* 117, 317–20.

SOMERS, I. I. and SHIVE, J. W. (1942). The iron–manganese relation in plant metabolism. *Plant Physiol.* 17, 582–602.

SOMMER, A. L. (1926). Studies concerning the essential nature of aluminium and silicon for plant growth. *Univ. Calif. Publ. Agric. Sci.* 5, 57–81.

SOMMER, A. L. (1928). Further evidences of the essential nature of zinc for the growth of higher green plants. *Plant Physiol.* 3, 217–21.

SOMMER, A. L. (1931). Copper as an essential for plant growth. *Plant Physiol.* 6, 339–45.

SOMMER, A. L. and BAXTER, A. (1942). Differences in growth limitation of certain plants by magnesium and minor element deficiencies. *Plant Physiol.* 17, 109–15.

SOMMER, A. L. and LIPMAN, C. B. (1926). Evidence of the indispensable nature of zinc and boron for higher green plants. *Plant Physiol.* 1, 231–49.

STEENJBERG, F. (1950). Investigations on micro-elements from a practical point of view. In 'Trace Elements in Plant Physiology'. *Lotsya*, 3, 87–97.

STEINBECK, O. (1951). Untersuchungen über Bormangelerscheinungen bei Kartoffeln. *Bodenkultur*, **5**, 57–60.

STEINBERG, R. A. (1919). A study of some factors in the chemical stimulation of the growth of *Aspergillus niger*. *Amer. J. Bot.* **6**, 330–72.

STEINBERG, R. A. (1935a). The nutritional requirements of the fungus *Aspergillus niger*. *Bull. Torrey Bot. Cl.* **62**, 81–90.

STEINBERG, R. A. (1935b). Nutrient-solution purification for removal of heavy metals in deficiency investigations with *Aspergillus niger*. *J. Agric. Res.* **51**, 413–24.

STEINBERG, R. A. (1936). Relation of accessory growth substances to heavy metals including molybdenum, in the nutrition of *Aspergillus niger*. *J. Agric. Res.* **52**, 439–48.

STEINBERG, R. A. (1937). Role of molybdenum in utilization of ammonium- and nitrate-nitrogen by *Aspergillus niger*. *J. Agric. Res.* **55**, 891–902.

STEINBERG, R. A. (1938a). Applicability of nutrient-solution purification to the study of trace-element requirements of *Rhizobium* and *Azotobacter*. *J. Agric. Res.* **57**, 461–76.

STEINBERG, R. A. (1938b). The essentiality of gallium to growth and reproduction of *Aspergillus niger*. *J. Agric. Res.* **57**, 569–74.

STEINBERG, R. A. (1938c). Correlations between biological essentiality and atomic structure of the chemical elements. *J. Agric. Res.* **57**, 851–8.

STEINBERG, R. A. (1939). Growth of fungi in synthetic nutrient solutions. *Bot. Rev.* **5**, 327–50.

STEINBERG, R. A. (1941). Use of *Lemna* for nutrition studies on green plants. *J. Agric. Res.* **62**, 423–30.

STEINBERG, R. A. (1942). Influence of carbon dioxide on response of *Aspergillus niger* to trace elements. *Plant Physiol.* **17**, 129–32.

STEINBERG, R. A. (1948). Essentiality of calcium in the nutrition of fungi. *Science*, **107**, 423.

STEINBERG, R. A. (1955). Effect of boron deficiency on nicotine formation in tobacco. *Plant Physiol.* **30**, 84–6.

STEINBERG, R. A. and JEFFREY, R. N. (1956). Effect of micronutrient deficiencies on nicotine formation by tobacco in water culture. *Plant Physiol.* **31**, 377–82.

STEINBERG, R. A., SPECHT, A. W. and ROLLER, E. M. (1955). Effect of micronutrient deficiencies on mineral composition, nitrogen fractions, ascorbic acid and burn of tobacco grown to flowering in water culture. *Plant Physiol.* **30**, 123–9.

STEWART, I. and LEONARD, C. D. (1952). Molybdenum deficiency in Florida *Citrus*. *Nature, Lond.*, **170**, 714–15.

STEWART, I. and LEONARD, C. D. (1953). Correction of molybdenum deficiency in Florida *Citrus*. *Proc. Amer. Soc. Hort. Sci.* **62**, 111–15.

STEWART, J., MITCHELL, R. L. and STEWART, A. B. (1941). Pining in sheep: its control by administration of cobalt and by use of cobalt-rich fertilizers. *Empire J. Exp. Agric.* **9**, 145–52.

STOKLASA, J. (1911). De l'importance physiologique du manganèse et de l'aluminium dans la cellule végétale. *C.R. Acad. Sci., Paris*, **152**, 1340.

STOKLASA, J. (1922). *Über die Verbreitung des Aluminums in der Natur.* Jena.

STOREY, H. H. and LEACH, R. (1933). A sulphur deficiency disease of the tea bush. *Ann. Appl. Biol.* **20**, 23–56.

STORP, F. (1883). Ueber den Einfluss von Kochsalz- und Zincsulfathaltigen Wasser auf Boden und Pflanzen. *Landw. J.* **12**, 795–844.

STOUT, P. R. and ARNON, D. I. (1939). Experimental methods for the study of the role of copper, manganese, and zinc in the nutrition of higher plants. *Amer. J. Bot.* **26**, 144–9.

STOUT, P. R. and JOHNSON, C. M. (1956). Molybdenum deficiency in horticultural and field crops. *Soil Sci.* **81**, 183–90.

STOUT, P. R., LEVY, J. and WILLIAMS, L. C. (1938). Polarographic studies with the dropping mercury kathode. Part LXXIII. The estimation of zinc in the presence of nickel, cobalt, cadmium, lead, copper and bismuth. *Coll. Czechoslovak Chem. Comm.* **10**, 129–35.

STOUT, P. R. and MEAGHER, W. R. (1948). Studies of the molybdenum nutrition of plants with radioactive molybdenum. *Science,* **108**, 471–3.

STROUTS, C. R. N., GILFILLAN, J. H. and WILSON, H. N. (editors) (1955). *Analytical Chemistry. The Working Tools.* Oxford.

STUMPF, P. K. and LOOMIS, W. D. (1950). Observations on a plant amide enzyme system requiring manganese and phosphate. *Arch. Biochem.* **25**, 451–3.

STUMPF, P. K., LOOMIS, W. D. and MITCHELSON, C. (1951). Amide metabolism in higher plants. I. Preparation and properties of a glutamyl transphorase from pumpkin seedlings. *Arch. Biochem.* **30**, 126–37.

SWANBACK, T. R. (1927). The effect of boric acid on the growth of tobacco plants in nutrient solutions. *Plant Physiol.* **2**, 475–86.

SWANBACK, T. R. (1939). Studies on antagonistic phenomena and cation absorption in tobacco in the presence and absence of manganese and boron. *Plant Physiol.* **14**, 423–46.

TALIBLI, G. A. (1935). Bedeutung von Mikroelementen und des Verhältnisses von Ca/Mg für das Pflanzenwachstum bei Kalkungen säurer Boden. *Z. Pflanzenernähr., Düng. u. Bodenk.* A, **39**, 257–64.

TATE, F. G. H. and WHALLEY, H. K. (1940). The spectrographic analysis of tobacco ash. *Analyst,* **65**, 587–93.

THACKER, E. J. and BEESON, K. C. (1958). Occurrence of mineral deficiencies and toxicities in animals in the United States and problems of their detection. *Soil Sci.* **85**, 87–94.

THATCHER, R. W. (1934). A proposed classification of the chemical elements with respect to their function in plant nutrition. *Science,* **79**, 463–6.

THEORELL, H. and SWEDIN, B. (1939). Mangan als Aktivator der Dioxymaleinsäureoxydase. *Naturwiss.* **27**, 95.

THOMAS, B. and TRINDER, N. (1947). The ash components of some moorland plants. *Emp. J. Exp. Agric.* **15**, 237–48.

THOMAS, H. E. (1931). The curing of exanthema by injection of copper sulphate into the tree. *Phytopathology*, **21**, 995–6.

TOTTINGHAM, W. E. and BECK, A. J. (1916). Antagonism between manganese and iron in the growth of wheat. *Plant World*, **19**, 359–70.

TREBOUX, O. (1903). Einige stoffliche Einflusse auf die Kohlensäureassimilation bei submersen Pflanzen. *Flora*, **92**, 56–8.

TRELEASE, S. F. and MARTIN, A. L. (1936). Plants made poisonous by selenium absorbed from the soil. *Bot. Rev.* **2**, 373–96.

TRELEASE, S. F. and TRELEASE, HELEN M. (1938). Selenium as a stimulating and possibly essential element for indicator plants. *Amer. J. Bot.* **25**, 372–80.

TRUE, R. H. and GIES, W. J. (1903). On the physiological action of some of the heavy metals in mixed solutions. *Bull. Torrey Bot. Cl.* **30**, 300–402.

TSUI, C. (1948). The role of zinc in auxin synthesis in the tomato plant. *Amer. J. Bot.* **35**, 172–9.

TWYMAN, E. S. (1943). Manganese deficiency in oats. *Nature, Lond.*, **152**, 216.

TWYMAN, E. S. (1951). The iron and manganese requirements of plants. *New Phytol.* **50**, 210–26.

TWYMAN, F. (1935). *The Practice of Spectrum Analysis with Hilger Instruments*. Sixth edition. London.

TWYMAN, F. (1938a). *Spectrochemical Abstracts*, 1933–7. London.

TWYMAN, F. (1938b). *Spectrochemical Analysis in* 1938. London.

TWYMAN, F. (1941). *The Spectrochemical Analysis of Metals and Alloys*. London.

ULRICH, A. and OHKI, K. (1956). Chlorine, bromine and sodium as nutrients for sugar beet. *Plant Physiol.* **31**, 171–81.

UNDENÄS, S. (1937). Ett försök med kopparsulfat mot gulspetsskuja. *Landboukshogskolano An.* (*Ann. Agric. Coll. Sweden*), **4**, 99–111.

UNDERHILL, F. P. and PETERMAN, F. I. (1929). Studies in the metabolism of aluminium. I. Method for determination of small amounts of aluminium in biological material. *Amer. J. Physiol.* **90**, 1–14.

UNDERWOOD, E. J. and FILMER, J. F. (1935). Enzootic marasmus. The determination of the biologically potent element (cobalt) in limonite. *Aust. Vet. J.* **11**, 84–91.

UNDERWOOD, E. J. and HARVEY, R. J. (1938). Enzootic marasmus: the cobalt content of soils, pastures and animal organs. *Aust. Vet. J.* **14**, 183–9.

VALLEE, B. L. and HOCH, F. L. (1955). Yeast alcohol dehydrogenase, a zinc metalloenzyme. *J. Amer. Chem. Soc.* **77**, 821.

VALLEE, B. L., HOCH, F. L., ADELSTEIN, S. J. and WACKER, W. E. L. (1956). Pyridine-nucleotide-dependent metallo-dehydrogenases. *J. Amer. Chem. Soc.* **78**, 5879–83.

VAN SCHREVEN, D. A. (1934). Uitwendige en inwendige symptomen van boriumgebrek bij tabak. *Tijdschr. PlZiekt.* **40**, 98–129 (with English summary).

VAN SCHREVEN, D. A. (1935). Uitwendige en inwendige symptomen van boriumgebrek bij tomaat. *Tijdschr. PlZiekt.* **41**, 1–26 (with English summary).

VAN SCHREVEN, D. A. (1939). De gezondheidstoestand van de aardappelplant onder den invloed van twaalfelementen. *Meded. Inst. Phytopath. Wageningen*, **43**, 166 pp. (with English summary).

VANSELOW, A. P. and BRADFORD, G. R. (1957). Techniques and applications of spectroscopy in plant nutrition studies. *Soil Sci.* **83**, 75–83.

VANSELOW, A. P. and DATTA, N. P. (1949). Molybdenum deficiency of the *Citrus* plant. *Soil Sci.* **67**, 363–75.

VANSELOW, A. P. and LAURANCE, B. M. (1936). Spectrographic micro-determination of zinc. *Industr. Engng Chem.* (Anal. ed.), **8**, 240–2.

VIETS, F. G., BOAWN, L. C. and CRAWFORD, C. L. (1954). Zinc content of bean plants in relation to deficiency symptoms. *Plant Physiol.* **29**, 76–9.

VINOGRADOV, A. P. (1934). Distribution of vanadium in organisms. *C.R. Acad. Sci. U.R.S.S.* pp. 454–9 (with English summary).

WADLEIGH, C. H. and SHIVE, J. W. (1939). A microchemical study of the effect of boron deficiency in cotton seedlings. *Soil Sci.* **47**, 33–6.

WAHHAB, A. and BHATTI, H. M. (1958). Trace element status of some West Pakistan soils. *Soil Sci.* **86**, 319–23.

WALKER, J. B. (1953). Inorganic micronutrient requirements of *Chlorella*. I. Requirements for calcium (or strontium) copper and molybdenum. *Arch. Biochem. Biophys.* **46**, 1–11.

WALKER, J. B. (1954). Inorganic micronutrient requirements of *Chlorella*. II. Quantitative requirements for iron, manganese and zinc. *Arch. Biochem. Biophys.* **53**, 1–8.

WALKER, J. C. (1939). Internal black spot of garden beet. *Phytopathology*, **29**, 120–8.

WALKER, J. C., JOLIVETTE, J. P. and McLEAN, J. C. (1943). Boron deficiency in garden and sugar beet. *J. Agric. Res.* **66**, 97–123.

WALKER, J. C., McLEAN, J. G. and JOLIVETTE, J. P. (1941). The boron deficiency disease in cabbage. *J. Agric. Res.* **62**, 573–87.

WALKER, T. W., ADAMS, A. F. R. and ORCHISTON, H. D. (1955). The effects and interactions of molybdenum, lime and phosphate treatments on the yield and composition of white clover, grown on acid, molybdenum-responsive soils. *Plant and Soil*, **6**, 201–20.

WALKLEY, A. (1942). The determination of zinc in plant materials. *Aust. J. Exp. Biol. Med. Sci.* **20**, 139–47.

WALLACE, A. and BEAR, F. E. (1949). Influence of potassium and boron on nutrient element balance in the growth of ranger alfalfa. *Plant Physiol.* **24**, 664–80.

WALLACE, T. (1943). *The Diagnosis of Mineral Deficiencies in Plants.* London. (Supplement, 1944.)

WALLACE, T. and HEWITT, E. J. (1946). Studies in iron deficiency of crops. I. Problems of iron deficiency and the interrelationships of mineral elements in iron nutrition. *J. Pomol.* **22**, 153–61.

WALLACE, T., HEWITT, E. J. and NICHOLAS, D. J. D. (1945). The resolution of factors injurious to plants on acid soils. *Nature, Lond.*, 156, 778.

WARINGTON, K. (1923). The effect of boric acid and borax on the broad bean and certain other plants. *Ann. Bot., Lond.*, 37, 629–72.

WARINGTON, K. (1926). The changes induced in the anatomical structure of *Vicia faba* by the absence of boron from the nutrient solution. *Ann. Bot., Lond.*, 40, 27–42.

WARINGTON, K. (1934). Studies in the absorption of calcium from nutrient solutions with special reference to the presence or absence of boron. *Ann. Bot., Lond.*, 48, 743–76.

WARINGTON, K. (1937). Observations on the effect of molybdenum in plants with special reference to the Solanaceae. *Ann. Appl. Biol.* 24, 475–93.

WARINGTON, K. (1940). The growth and anatomical structure of the carrot (*Daucus carota*) as affected by boron deficiency. *Ann. Appl. Biol.* 27, 176–83.

WAYGOOD, E. R. and CLENDENNING, K. A. (1950). Carbonic anhydrase in green plants. *Canad. J. Res.* C, 28, 673–89.

WAYGOOD, E. R., OAKS, A. and MacLACHLAN, C. A. (1956a). On the mechanism of indoleacetic acid oxidation by wheat leaves. *Canad. J. Bot.* 34, 54–9.

WAYGOOD, E. R., OAKS, A. and MacLACHLAN, C. A. (1956b). The enzymatically catalysed oxidation of indoleacetic acid. *Canad. J. Bot.* 34, 905–26.

WEAR, J. I. (1956). Effect of soil pH and calcium on uptake of zinc by plants. *Soil Sci.* 81, 311–15.

WEINBERGER, J. H. and CULLINAN, F. P. (1937). Symptoms of some mineral deficiencies in one-year Elberta peach trees. *Proc. Amer. Soc. Hort. Sci.* 1936, 34, 249–54.

WEINSTEIN, L. H. and ROBBINS, W. R. (1955). The effect of different iron and manganese nutrient levels on the catalase and cytochrome oxidase activities of green and albino sunflower leaf tissues. *Plant Physiol.* 30, 27–32.

WHITEHEAD, T. (1935). A note on 'brown-heart', a new disease of swedes, and its control. *Welsh J. Agric.* 11, 235–6.

WICKENS, G. W. (1925). Exanthema of *Citrus* trees. *Rep. Proc. Imp. Bot. Conference*, London, 1924, pp. 353–7. Cambridge.

WIESE, A. C. and JOHNSON, B. C. (1939). Microdetermination of manganese. *J. Biol. Chem.* 127, 203–9.

WIKLANDER, L. (1958). The soil. In *Encyclopedia of Plant Physiology*. Vol. IV, pp. 118–69. Berlin, Göttingen, Heidelberg.

WILCOX, L. V. (1940). Determination of boron in plant material. An ignition-electrometric titration method. *Industr. Engng Chem.* (Anal. ed.), 12, 341–3.

WILLIAMS, D. E. and VLAMIS, J. (1957). Manganese toxicity in standard culture solutions. *Plant and Soil*, 8, 183–93.

WILSON, R. D. (1949). Molybdenum in relation to the scald disease of beans. *Aust. J. Sci.* 11, 209–11.

WINFIELD, M. E. (1945). The role of boron in plant metabolism. III. The influence of boron on certain enzyme systems. *Aust. J. Exp. Biol. Med. Sci.* **23**, 267–72.

WOLFF, L. K. and EMMERIE, A. (1930). Über das Wachstum des *Aspergillus niger* und den Kupfergehalt des Nährbodens. *Biochem. Z.* **228**, 441–50.

WOODWARD, J. (1699). Some thoughts and experiments concerning vegetation. *Philos. Trans.* **21**, 193–227.

WOOLEY, D. W. (1941). Manganese and the growth of lactic acid bacteria. *J. Biol. Chem.* **140**, 311–12.

WUNSH, D. S. (1937). Tracking down a deficiency disease. *Chem. and Ind.* **15**, 855–9.

YAKOVLEVA, V. V. (1947). The influence of boron on the biochemical changes in the roots and leaves of the sugar beet. *Dokl. Akad. Nauk U.R.S.S.* **58**, 625–7. Cited from *Chem. Abstr.* **45**, 8091.

YARWOOD, C. E. (1942). Stimulatory and toxic effects of copper sprays on powdery mildews. *Amer. J. Bot.* **29**, 132–5.

YOE, J. H. and WILL, F. (1952). A new colorimetric reagent for molybdenum. *Anal. chim. acta*, **6**, 450–1.

YOUNG, R. S. (1935). Certain rarer elements in soils and fertilizers and their role in plant growth. *Mem. Cornell Agric. Exp. Sta.* no. 174, 70 pp.

INDEX

Abraham, R. G., 150
absorptiometer, 22
absorptiometric determination
    of aluminium, 50
    of boron, 46–7
    of cobalt, 52
    of copper, 42–4
    of manganese, 33–5
    of molybdenum, 48
    of zinc, 39–41
absorption
    of manganese, retarded by calcium,
        125–6
    of selenium by plants, 182–4
    of trace elements, factors in-
        fluencing, 114–42
Acetobacter aceti, effect of copper on
    oxidation by, 163–4
Adams, A. F. R., 168
Adams, M. B., 119, 138
Adelstein, S. J., 155
Agaricus campestris, polyphenol oxi-
    dase in, 161
Agulhon, H., 3
Alben, A. O., 73
alcohol dehydrogenase, 156
aldolase, 156
Aleurites fordii, see tung tree
Aleurites montana, see mu-oil tree
Alexander, T. R., 84–5
alfalfa, *see* lucerne
algae, trace elements necessary for,
    2, 9–11
alkali disease, 180–4
alkaline fen soil, copper content of,
    137
Allen, P. J., 156
Allison, F. E., 167
almond, zinc deficiency in, 75
aluminium
    estimation of, in ash and soil, 49–51
    species requiring, 4–5
    as micro-nutrient, 3–5
Amin, J. V., 140
amino acids in plants and molybdenum
    supply, 165, 168
ammonium salts, effect of, on uptake
    of manganese, 126
Anabaena variabilis, sodium essential
    for, 111

Anacystis nidulens, sodium essential
    for, 11
Anderson, A. J., 167–8
Anderson, I., 152
Anderssen, F. G., 6, 54, 57, 97–8
Andropogon scoparius, selenium con-
    tent of, 182
Andropogon sorghum, see Sorghum
animals, grazing, trace elements in
    plants in relation to, 180–94
anions, effect of boron on absorption
    of, 170–1
antagonism between manganese and
    calcium, 125–6
apical buds, effect of zinc deficiency
    on, 71–2
apple
    boron deficiency in, 92
    copper deficiency in, 96–7
    internal cork in, 92
    intervenal injection of, 55
    leaf-stalk injection of, 56
    terminal oxidase in, 162
    zinc deficiency in, 75
apricot
    boron deficiency in, 93
    brown spotting of, 93
    zinc deficiency in, 70–2, 75–6
arc spectra, 25–7
Argawala, S. C., 165–6
Ark, P. A., 76, 196
Arnon, D. I., 1, 6, 10–12, 19–21, 101,
    147, 167
ascorbic acid, 161, 166
    oxidase, 160–2
ash, preparation of, 30–1, 43
Askew, H. O., 52, 92–3, 190
Aso, K., 106–7, 119
asparagus, effect of trace elements on
    growth of, 12–13
Aspergillus flavus, micronutrients
    for, 8
Aspergillus niger
    copper essential for, 8
    effect of micronutrients on growth
        of, 7–8, 148
    manganese and metabolism of, 152
    micronutrients for, 7
    molybdenum enzyme in, 164;
        essential for, 8, 164

maize (*cont.*)
  chlorine: deficiency in, 105; necessary for, 3, 105
  copper–boron relations in, 162
  iron–manganese relations in, 120
  manganese: deficiency in, 62–3; excess in, 108
  micro-nutrients for, 3
  white bud of, 3, 76, 81–3
  zinc: deficiency in, 70–1, 76, 81–3; necessary for, 3
Majdel, J., 34
malic dehydrogenase, 149
malic enzyme, 151–2
manganese
  availability, effect of soil conditions on, 5–17
  concentration, effect of, on growth of leguminous plants, 108
  content: of bracken, 118; of dwarf bean plants, 108–9; of grasses, 119; of herbaceous plants, 129; of lichens, 128; of *Molinia coerulea*, 118–19; of oats, 60; of para grass, 119; of peas, 66; of plants, 118–19; of plum tree leaves, 98; of red top, 119; of rose-bay willow-herb, 118; of *Scirpus caespitosus*, 118–19; of sheep sorrel, 118; of soils, 117–18; of tobacco plant, 126; of sugar beet, 65; of *Vaccinium myrtillus*, 118–19
  deficiency diseases, 58–69: in barley, 62; in broad bean, 68; in dwarf bean, 68; in garden beet, 64; in maize, 62–3; in mangold, 64; in mu-oil tree, 68; in oat, 59–61; in peas, 65–8; in runner bean, 68; in rye, 62; in spinach, 64; in sugar beet, 64; in sugar cane, 63–4; in tung, 68–9; in wheat, 62
  determination of, in plant ash, 31–5
  in enzyme actions, 148–54
  excess: in barley, 106–7; in cauliflower, 107; in *Citrus*, 106; in clover, 107; in cowpea, 107; in dwarf bean, 106; in *Lemna polyrhiza*, 108; in *Lespedesa*, 107; in maize, 108; in oats, 107; in peas, 107; in plants, 106–9; in potato, 107; in *Senecio sylvatica*, 108; in soya bean, 107; in strawberry, 107; in tobacco, 107; in vetch, 106; in wheat, 106–7

functions of, in plants, 143, 145–55
  as micro-nutrient, 2–3, 7–8
  in soil, 115–19
manganese–iron relations in plants, 119–24, 154–5, 163
manganese supply, effect of, on yield of peas, 68
mangold
  boron deficiency in, 86–7
  heart rot of, 86–7
  manganese deficiency in, 64
Mann, M., 7
Mann, T., 155, 161
marasmus, enzootic, 189–94
Marloth, R. H., 7, 9
Marmoy, F. B., 48
marrowstem kale, trace element interactions in, 197
Marsh, R. P., 173–4, 176
marsh spot, 65–8, 153
Marston, H. R., 52
Martin, A. L., 184
Martin, D., 92
Martin, J. P., 41, 119, 131, 133, 138, 144
Maschhaupt, J. G., 116
Massey, D. M., 119
Matthews, E. M., 91
Maunsell, P. W., 191
Mazé, P., 3–4, 13, 81, 105, 158
Meagher, W. R., 198
Mederski, H. J., 107, 111, 126
Medicago sativa, *see* lucerne
Medina, A., 156
Mehta, B. V., 127
Meiklejohn, G. T., 162
Melampsora lini, protective action of boron against, 179
Meldrum, N. U., 157
Melvin, E. H., 31, 41, 45
Menzel, R. G., 38, 42
Metz, O., 9
micro-organisms associated with: grey speck, 60–1; little-leaf and white bud, 76–7
microphotometer, 27
Miller, C. E., 136
Miller, E. J., 39
Miller, J. T., 182
Miller, W. T., 181–2
millet
  aluminium necessary for, 5
  boron: excess in, 111–12; uptake by, 136

Printed in the United States
By Bookmasters